D0952869

MATH
BYTES

TIM CHARTIER

GOOGLE BOMBS,
CHOCOLATE-COVERED PI,
AND OTHER COOL
BITS IN COMPUTING

MATH BYTES

PRINCETON UNIVERSITY PRESS
PRINCETON AND OXFORD

Copyright © 2014 by Princeton University Press
Published by Princeton University Press, 41 William Street,
Princeton, New Jersey 08540
In the United Kingdom: Princeton University Press,
6 Oxford Street, Woodstock, Oxfordshire, OX20 1TW

press.princeton.edu

All Rights Reserved
Library of Congress Cataloging-in-Publication Data

Chartier, Timothy P., 1969–
Math bytes : Google bombs, chocolate-covered pi, and other cool
bits in computing / Timothy P. Chartier.
pages cm
Includes bibliographical references and index.
ISBN 978-0-691-16060-3 (acid-free paper) 1. Mathematics–Popular works.
2. Mathematics–Data processing–Popular works.
3. Computer science–Popular works. I. Title.
QA93.C47 2014
510–dc23

2013025378

British Library Cataloging-in-Publication Data is available

This book has been composed in Minion Pro

Printed on acid-free paper ∞

Typeset by S R Nova Pvt Ltd, Bangalore, India
Printed in the United States of America

1 3 5 7 9 10 8 6 4 2

To Noah and Mikayla,
snuggle up close,
I have math-nificent stories to share
before this day ends,
and then may your dreams lead you
to explore what you learn in new ways.
I await your stories. . . .

CONTENTS · · · · · · · · · · · · · · · · · ·

PREFACE ●

"WHY STUDY MATH?" can be a common question about mathematics. While a variety of answers are possible, seeing applications of math and how it relates to the world can be inspiring, motivating, and eye opening. This book discusses mathematical techniques that can recognize a disguised celebrity and rank web pages with the roll of a pair of dice. Math also helps us understand our world. How can poor reviews on Twitter impact a big budget film's poor showing on opening weekend? How are fonts created by computers?

Mathematics has many applications, too many for any book. The purpose here is to cover and introduce mathematical ideas that use computing. Some of the ideas were developed by others, like Google's PageRank algorithm. Other ideas were developed specifically for this book, like how to create mazes from mathematical images created with TSP Art. The goal is to give the reader a good taste for computing and mathematics (thus a *byte*). Still, many ideas are not presented in their full depth with the mathematical theory or most complex (and robust) approach (thus a *bit* of math and computing).

This book is intended for broad audiences. It was developed in this way. For example, Chapter 6 begins with math and doodling, which is adapted from content taught to non-majors at Davidson College. Later in the chapter, the content on the Traveling Salesman Problem and TSP Art was developed from lectures to math majors at Davidson College and the University of Washington. The chapter concludes with the creation of mazes, which is an application of these ideas developed specifically for this book. Other chapters have been adapted from ideas developed in my seminars with the Charlotte Teachers Institute. As such, teachers should find ideas that align with their curricular goals.

Some sections of this book have been used in my public presentations in which audiences range from children to adults. Other sections, like the later portions of Chapter 12 on the theory of PageRank, tend to appeal to mathematicians. This multileveled approach is intentional. Given the eclectic collection of ideas, I imagine a reader can delve into what's interesting and skim sections of less interest. Throughout the development of this book, readers have ranged from self-proclaimed "math haters" to professors of mathematics. Wherever you, as a reader, fall in this range, may you enjoy these applications of mathematics and computing.

Throughout the book, you'll find hands-on activities from using chocolate to estimating π to computing a trajectory of flight in Angry Birds. I expect such explorations to be a tasty byte of math and computing for many readers. If you enjoy these investigations, I encourage you to view the supplemental resources for this book at the Princeton University Press website; you'll find such materials at http://press.princeton.edu/titles/10198.html.

My hope is that this book will broaden a reader's perspective of how math can be used. So, when someone turns to you and asks for a reason to study math, you'll at least have one, if not several, reasons. Math is used every day. Such applications can be enjoyed and appreciated even if you simply consider a bit and a byte of it.

Acknowledgments. First and foremost, I thank Tanya Chartier who walks this journey with me and affirms choices that lead to roads less travelled and sometimes possibly even untrodden. Tanya encouraged me to write the first draft of this book when it was clearest in my mind during my sabbatical at the University of Washington in Seattle. Thank you, Tanya, for such wise counsel. I'm grateful to Melody Chartier for being my first reader as her input supplied great momentum and the ever-present goal of reaching a broad audience. I also thank Ron Taylor, Brian McGue, and Daniel Orr for their readings of the book. I enjoyed developing the Jay Limo story and chocolate Calculus with Austin Totty. I'm grateful for the various lunch conversations with Anne Greenbaum that largely led to our coauthored text [18] but also inspired a variety of ideas for this book. I appreciate Davidson College's

Alan Michael Parker for encouraging me, when I thought I was done with an earlier draft, to write "my book" with "my voice." Having studied mime with Marcel Marceau while I pursued my doctorate in Applied Mathematics, it seems almost natural to have written a book that computes π with chocolate, finds math in a popular video game, and explores math in sports. As such, I believe this book reflects my journey as a mathematician and artist. I thank my parents who taught me to apply my learning beyond preset boundaries and to look for connections in unexpected areas. Finally, I appreciate the many ways Vickie Kearn has supported me in this writing process and her guidance, friendship, and encouragement. It is difficult, if not impossible, to thank everyone who played a part in this book. To colleagues, students, and friends, may you see your encouragement and the results of conversations in these pages.

**MATH
BYTES**

1. . . . .

Your First Byte

MATHEMATICS HELPS CREATE THE LANDSCAPES on distant planets in the movies. The web pages listed by a search engine are possible through mathematical computation. The fonts we use in word processors result from graphs of functions. These applications use a computer although each is possible to perform by hand, at least theoretically. Modern computers have allowed applications of math to become a seamless part of everyday life.

Before the advent of the digital computer, the only computer we had to rely on was a person's mind. Few people possessed the skill to perform complex computations quickly. Johan Zacharias Dase (1824–1861) was noted for such skill and could mentally multiply

$$79{,}532{,}853 \times 93{,}758{,}479 = 7{,}456{,}879{,}327{,}810{,}587$$

in 54 seconds, two 20-digit numbers in six minutes, two 40-digit numbers in 40 minutes, and two 100-digit numbers in 8 hours and 45 minutes. As another example, Leonhard Euler, the most prolific mathematical writer of all time with over 800 published papers, was renowned for his phenomenal memory. Such ability became important during the productive last two decades of his life, when he was totally blind. He once performed a calculation in his head to settle an argument between students whose computations differed in the fiftieth decimal place.

Multiprocessors, in a certain sense, also existed. Gaspard Riche de Prony calculated massive logarithmic and trigonometric tables by employing systematic division of mental labor. In one step of the process, mathematically untrained hairdressers unemployed after the French Revolution performed only additions and subtractions. De Prony's process of calculation inspired Charles Babbage in the design of his difference engine, which was an automatic, mechanical calculator operated by the turn of a crank. The advent of computational machinery added a level of certainty in comparison to hand calculations. Computing machines could benefit from parallel calculations, seen in Figure 1.1, much like de Prony's calculations did from his team of hairdressers.

Algorithmic advances play a key role in the efficiency of modern mathematical computing. Consider simulating a galaxy. Prior to the

Figure 1.1. Early parallel computing? At least 30 workers at the Computing Division, Veterans Bureau, in Washington, DC in the early 1920s used Burroughs electric adding machines to compute bonuses for World War I veterans.

1970s, a computer simulating a cluster of 1,000 stars would compute the attraction of each pair of stars on each other resulting in 999,000 computations. The motion of a star is determined by summing the 999 forces tugging on it by other stars. If a computer calculates this for all 1,000 stars for many instances in time, the motion of the cluster emerges and we see how that galaxy evolves over time.

In the 1970s, a cheaper computation was discovered. The process begins by pairing each star with its nearest neighbor, then grouping each pair with its closest pair, and subsequently clustering each superpair with its nearest neighbor until a single group remains. The forces are then computed for the groupings on each level. For our galaxy of 1,000 stars this results in a simulation that involves 10,000 computations at each step in time.

SUPERCOMPUTING AND LINEAR SYSTEMS

The U. S. national laboratories house many of the world's fastest high performance computers. Such computers enable scientists to simulate complex physical phenomena, sometimes involving billions of variables. One of the fastest computers in the world (in fact the fastest in June 2010) is Jaguar (pictured above) housed at Oak Ridge National Laboratory that can perform 2.3 quadrillion floating-point operations per second (flops). A flop is a measure of computer performance measuring the number of floating-point calculations performed per second, similar to the older, simpler instructions per second.

Algorithmic improvements increase the speed of computation. Improvements in processor speed make things even faster. Before 1970, computers performed less than 1 million instructions per second (MIPS). Today, laptops easily perform over 1,000 times as fast. The Intel⑧Core i7-990X chip, released in 2011, operates at 3.46 GHz and reaches 159,000 MIPS. This means what can be done in a second or less on a computer today would have taken over 16 minutes by that 1975 mainframe. However, this assumes we run the same algorithm on both computers. Suppose we run our modern algorithm (which involves 1/100 as many computations) and its slower counterpart on the older mainframe. Now a simulation that takes 6 seconds of computation on today's computer would have taken over 2 months of constant computation on the 1975 mainframe.

These types of advances have led to the integral role of computing in our world, from ATMs to digital watches to home computers with internet browsers and word processing. In this book, we'll apply mathematics to a variety of topics. Computing will play an important role, sometimes making the application possible. In reading this book, you should get a taste of how math and computing influences our world and culture.

2.................

Deceiving Arithmetic

ADDITION AND SUBTRACTION CAN SEEM EASY enough. In fact, that is largely what a computer does, and a computer only adds 1s and 0s. Yet, even this most foundational mathematical operation can be tricky and require attention.

Homer Simpson's Method

In the 7th season of *The Simpsons*, Homer has a nightmare in which he travels into 3D space. In this episode entitled "Treehouse of Horror VI" as Homer struggles with this newfound reality, an equation flies past 3D Homer that reads:

$$1782^{12} + 1841^{12} = 1972^{12}.$$

Simple enough, unless you happen to remember some fateful words written in 1637.

It may have been a dark and dreary night in 1637 as Pierre de Fermat delved deeply into the work *Arithmetica* written by Diophantus some 2,000 years prior. Mathematical ideas swirled in Fermat's mind until suddenly the numbers fell into an unmistakable, miraculous pattern.

Quickly, he scratched into the margin of his copy of the ancient work the statement that would be forever linked with his name:

> It is impossible to write a cube as a sum of two cubes, a fourth power as a sum of two fourth powers, and, in general, any power beyond the second as a sum of two similar powers. For this, I have discovered a truly wondrous proof, but . . . the margin is too small to contain it.

Let's look at this statement carefully as it will help us understand the humor and ingenuity of the equation in Homer's 3D world. For any two positive integers, if we take the first number raised to the third power, then add it to the second number raised to the third power it will never equal a positive integer raised to the third power. Written mathematically:

$$x^3 + y^3 \neq z^3.$$

For instance,

$$2^3 + 3^3 = 8 + 27 = 35,$$

but the cube root of 35 is about 3.27, which isn't an integer. Fermat's statement became known as his Last Theorem and is stated mathematically as:

> **Fermat's Last Theorem** For $n \geq 3$, there do not exist any natural numbers x, y and z that satisfy the equation
>
> $$x^n + y^n = z^n.$$

What did Fermat actually "prove"? He was notorious for making mathematical claims with little or no justification or proof. By the 1800s, all of Fermat's statements had been resolved except the last one written on that fateful night.

On another night in 1994 in New Jersey, Princeton University professor Andrew Wiles discovered a proof of Fermat's Last Theorem. What kind of proof? It called on mathematics that had not yet been discovered in Fermat's time. He didn't write it in the margin . . . it took 130 pages.

Many brilliant minds worked on this problem and many mathematical advances took place even with the failed attempts. As more mathematicians failed to prove and disprove Fermat's statement, the "theorem" grew in its prestige. It turns out the initial proof even had an error in it—that was corrected. It took 357 years, but Fermat's Last Theorem was finally indeed proven to be a theorem. The accomplishment was even headline news, which is particularly newsworthy when the topic is mathematical.

Now, back to Homer's dream. Did Homer dream up a counterexample to Fermat's Last Theorem? Again, his world proposed that

$$1782^{12} + 1841^{12} = 1922^{12}.$$

Will there be new headlines regarding Homer's disproving Wiles' accomplishment? Let's wait a moment and look carefully at this equation. Do you notice anything odd about it? Indeed, there is. The left-hand side of the equation is the sum of an even number (1782^{12}) and an odd number (1841^{12}), which will be an odd number, whatever it happens to be. What about the right-hand side of the equation? It will be even since it is an even number raised to a power. Even Homer (hopefully) would recognize that a number cannot be even and odd at the same time.

So, why include the equation? The humor is evident if you try testing the equation on some calculators, especially those available at the time the show originally aired. You can test the equation by noting if $1782^{12} + 1841^{12} = 1922^{12}$ then

$$\sqrt[12]{1782^{12} + 1841^{12}} = 1922.$$

So, what do you get? On many calculators and even some computers that use mathematical software, you will find that computing $\sqrt[12]{1782^{12} + 1841^{12}}$ results in an answer of 1922.0000000. What happened? It becomes clearer if you can print out an additional digit which uncovers that $\sqrt[12]{1782^{12} + 1841^{12}}$ equals 1921.99999996. Many calculators of the time would have indicated the equality we just saw.

Pretty clever! Where is all this math coming from in *The Simpsons*? It turns out that many of the writers have a mathematical background. David S. Cohen (a.k.a. David X. Cohen), writer for *The Simpsons* and head writer and executive producer of *Futurama*, graduated magna cum laude with a bachelor's degree in physics from Harvard University and a master's degree in computer science from UC Berkeley. Watch *The Simpsons* carefully and you'll see bits of math appearing. *Futurama* has even more references to mathematics. To read about math appearing in these shows, visit [19].

Now, it is your turn with a challenge problem. They are posed throughout the book with answers in the back. See how you do. If you get a good idea, you might want to jot it down in the margins of this book.

Figure 2.1. Equations like the one that appeared in Homer Simpson's 3D world are called Fermat near-misses since they come close to contradicting the famous Last Theorem.

> **Challenge 2.1.** *In a later episode entitled "The Wizard of Ever-green Terrace," Homer emulates Thomas Edison and writes on a blackboard*
>
> $$3987^{12} + 4365^{12} = 4472^{12}.$$
>
> *Prove this to be false.*

Losing Dollars

It doesn't just take round-off error for one to make computational errors. Incorrect assumptions can slip into even the simplest of problems—creating a major problem! Let's see this in a word problem adapted from the inventive and influential writings of Martin Gardner [17].

Suppose your entire block decides to have a yard sale over two days. The participating houses fill a table with DVDs and CDs. The first day, the DVDs sell 2 for $3, and CDs, 3 for $3. At the end of the day, 30 DVDs and 30 CDs sell for a grand total of $75. This inspires the neighbors to look for more DVDs and CDs enabling another impressive display of entertainment discs to be available on the second day of sales. However, on this day, the discs sell in any combination at 5 for $6. Amazingly, 60 discs sell, just like the first day, but now totaling $72. Where did the missing $3 go?

It's only $3. Who cares? Now, suppose Jay Limo, known for his large collection of cars, buys eight Rolls Royces at a price of two for $500,000

and eight BMW convertibles at four for $340,000. Jay can close the deal with a $2,680,000 cashier's check. Before the cars are delivered, Jay has a change of heart. If we use the reasoning from the previous problem, Jay could sell the cars at a price of six for $840,000 or $140,000 each. I'll take six Rolls Royces, please, which costs me only $840,000 even though they are worth two million dollars according to the original sale. Ah, clever! In that case, I must buy all the cars or no deal. Suppose I do! Now, I only need to pay Jay $2,240,000 for all sixteen cars. That's a loss of $440,000!

> **Challenge 2.2.** *Where did Jay Limo's math go wrong leading to a loss of $440,000? Where did all that money go? Dig around . . . play with the numbers.*

Can 2 = 1?

Clearly, 2 does not equal 1 but can we produce an argument that seems to imply this to be the case? Let's see. Math errors can be subtle. Famous mathematicians have found errors in their proofs. To get a sense of how this happens, here is a logic puzzle for you to try. The argument below clearly goes astray since by the end we claim $2 = 1$.

Start with an identity:	$a = b$
Multiply both sides by a:	$a^2 = ab$
Subtract b^2 from both sides:	$a^2 - b^2 = ab - b^2$
Factor both sides:	$(a - b)(a + b) = b(a - b)$
Cancel the common term:	$a + b = b$
But $a = b$, so:	$b + b = b$
Adding:	$2b = b$
Divide both sides by b:	$2 = 1$

> **Challenge 2.3.** *The argument above starts with a = b and ends with 2 = 1, where is the logical misstep?*

3. ·

Two by Two

SUPPOSE YOU COULD TAKE A MILLION DOLLARS TODAY or be given one cent today and twice as much as the day before for the next 30 days. So, if you took the second option, you'd take home a penny today, two pennies tomorrow and so forth. Which would you choose? The more lucrative answer will give insight on how a movie might bomb soon into its opening weekend.

A General Algorithm

Let's step back into Roman times. General Terentius returns home after one of his many victorious battles. Soon after entering the city, he requests an audience with the emperor, and he is warmly received. Upon the meeting, the emperor promises a place in the Senate befitting the soldier's dignity. The general, announcing his imminent retirement, indicates his desire to return home and forego stations of power. Caesar requests the amount of a reward to bestow in order to recognize the general's distinguished accomplishments. Terentius comments how his years of brave military service have left him poor and without fortune. The emperor asks his retiring soldier to state an adequate sum. Terentius requests a million denarii, which is a bold request especially as Caesar was not a generous man. The emperor promises riches but delays stating an amount until noon the next day. In that meeting, Yakov Perelman's book, *Mathematics Can Be Fun* [26], states Caesar's offer:

> *In my treasury there are 5 million brass coins worth a million denarii. Now listen carefully. Thou wilt go to my treasury, take one coin and bring it here. On the next day thou wilt go to the treasury again and take another coin worth twice the first and place it beside the first. On the third day thou wilt get a coin worth four times the first, on the fourth day eight times, on the fifth sixteen times, and so on. I shall order to have coins of the required value minted for thee every day. And so long as thou hast the strength, thou mayst take the coins out of my treasury. But thou must do it thyself, without any help. And when thou canst no longer lift the coin, stop. Our agreement will have ended then, but all the coins thou wilt have taken out will be thy reward.*

Terentius enthusiastically thanks the emperor for his generosity and accepts the offer, greedily imagining his forthcoming immense piles of money. Each brass coin weighed 5 grams. The first day, the soldier quickly snatched a single coin from the treasury. Continuing his daily pilgrimage, the general effortlessly lifted a coin weighing the same as 2 brass coins or only 10 grams on the second day. However, before the end of the second week, the mighty soldier struggled to return with his reward. The fourteenth day's prize required Terentius to carry a coin weighing $5 * \left(2^{13}\right) = 40,960$ grams or just over 90 pounds. The trips to the treasury lasted only a few more days. On day 18, Terentius struggled and successfully, with the aid of his spear as a lever, moved a coin just over 655 kilograms or 1,444 pounds. The next day's reward weighed over a ton and could not be returned. In all, Terentius received 262,143 coins - 1 from the first day, 2 from the second day, 4 from the third day and so forth until receiving 131,072 coins on his last successful trip. The miserly emperor gloated in his clever scheme rewarding his brave general with just over 5% of the requested amount.

Let's modernize the story. Suppose you could choose a room filled with pennies, nickels, dimes, or quarters and enter the room on the first day and take 1 coin. The next day, you could return and take double the

Figure 3.1. Could the dupondius be the fabled coin? It was, indeed, valued at 1/5 a denarius. However, it is unlikely it would have ever been as light as 5 grams, which is the weight of a modern United States minted nickel. For instance, the dupondius weighed over 100 grams in 211 BC.

number of coins you took the previous day until you could no longer carry the weight of the coins. Suppose you train and are as strong as Terentius. He struggled and succeeded in carrying 655 kg from the emperor's treasury. Suppose the max you can carry is 700 kg. Which room do you choose to maximize your reward?

Let's try pennies and nickels. A penny weights 2.5 grams. The first day, you easily take 2.5 grams. On the second day, you take 5 grams. On the 10th day, you carry 512 grams or just over a pound. Continuing in this way, on the 19th day, your muscles would bulge with 655.36 kg from 262,144 pennies. So, from the first to last day, you took 1, 2, 4, and 8 pennies until your final haul of just over a quarter of a million pennies. In all, you would have made $5,242.87.

How about nickels which weigh 5 grams each? Note that a nickel weights twice that of a penny. So your final trip would occur on day 18 with 131,072 nickels. From day 1 to day 18, you'd gain $13,107.15.

> **Challenge 3.1.** *Suppose you could choose between a room full of dimes or of quarters. Keep in mind that dimes weigh 2.2 grams and quarters 5.6 grams. Which room would you choose assuming you cannot carry more than 700 kg?*

From Hercules to Blockbusters

We just saw the power of doubling. Let's see this again in Greek mythology and then how such growth may help a film boom or bust at the box office. Hydra was a water beast that possessed many heads and would rise up from the murky waters of the swamp near Lerna and terrorize the countryside. Hercules met the creature in the second of his Twelve Labors. The beast was a difficult prey to kill as two heads would spring from any one that was cut.

Hercules found the hydra with 9 heads. Suppose he cut one, then two would arise. The creature now would have 10 heads. If Hercules then cut 3 heads in a mighty blow, 6 would arise giving it 13 heads. Hercules was joined on this labor by his nephew Iolaus. Suppose together they cut 20 heads. A hydra with 20 more heads than when they started would

Figure 3.2. Hercules, indeed, had a herculean task when he fought a Hydra.

be staring at them. So, how did Hercules defeat the beast? Rather than cut a head of the hydra, Iolaus held a torch to the headless tendons of the neck. The flames prevented the growth of replacement heads. This enabled Hercules to defeat the beast and move to his third labor.

Now, let's move from hydra to hype. Every summer audiences eagerly anticipate a heavily-promoted movie that, despite all the hype, fails at the box office in the opening weekend. Indeed, its only hope in sales is at rental kiosks when it appears a few months later.

Suppose 150 people buy the \$10 ticket for the midnight showing of an expected summer blockbuster. During the film, one person exits to the lobby and posts a witty, disapproving review on Twitter. The average Twitter user has 126 followers. [23] Assume two of them retweet this negative review. Suppose each of those retweets is retweeted by two friends which then again lead to retweets by two friends. How many times would this pattern need to be repeated for the bad review to by tweeted, which includes retweeting, by a million people?

This is similar to our earlier computations. We begin with 1 tweet, which leads to 2 retweets, 4 retweets and then 8 retweets as seen in Figure 3.3. Let's refer to this tweet passing through 3 levels of retweeting. We can know how many Twitter birds appear in Figure 3.3 with the formula $2^4 - 1$. In fact, if we extend this structure, called a tree, to have k levels, then the total number of blue birds would be $2^{k+1} - 1$. So after 3 levels, 15 people tweeted the poor review, $15 = 2^4 - 1$. If this tweet passed through 19 levels, then $2^{20} - 1 = 1,048,575$ people would retweet the poor review. Keep in mind this assumes that only 2 followers retweet and the average number of followers is 126. So, the tweet would be received by more than 100 million Twitter users, depending on the amount of overlap of followers through all the retweeting. If only 10% of the people who tweeted the bad review decided not to see the film, this would result in a million dollars in lost ticket sales. There are also those who read but decide not to retweet the post, so it may take far fewer levels for the movie to lose a million dollars in ticket sales.

Keep in mind that this is a hyped film so it is likely that more than 1 person would tweet a negative review. This leads us to our challenge question.

Figure 3.3. Retweets can grow quickly to spread the word!

Challenge 3.2. *Suppose rather than starting with 1 person's tweet, 40 people tweet bad reviews of the movie. How many levels now equate to a million people tweeting the poor reviews? As before, we will assume the number of people who retweet doubles for every level. So, the first level has 40 people and then 80 and so forth.*

TWEETING THE NEWS

In 2012, IBM Research produced summaries of sporting events from status updates posted to Twitter. [25] During a game, a variety of tweets describe and express opinions about the event. Analyzing the volume of tweets and ranking the importance of the tweeted content, the research group produced a computer algorithm that creates summaries of sporting events that is comparable to human-generated reports.

Tweets Galore

A video or topic going "viral" on the internet is another example of quick growth. From YouTube to Twitter, information can spread at alarming rates. Twitter is a social networking service with more than 500 million users worldwide as of 2012 generating over 340 million tweets daily, according to Wikipedia. Users tweet by sending a text-based post of up to 140 characters and read the tweets of the authors they choose to follow.

Not surprisingly, some topics are more tweeted than others. For instance, at 10:35 PM on Sunday, April 28, 2011, a surge of tweets appeared on Twitter posting a record 8,868 tweets per second on pop singer Beyoncé. She had just announced her pregnancy during her performance saying, "I want you to feel the love that's growing inside me." How much information is this?

While tweets cannot exceed 140 characters in length, their average length is 81.9 characters according to MediaFuturist. [23] So, in one second, 726,289 characters zoomed through Twitter. According to [27], the average length of a word (in English) is 5.5 characters. Note, here lies an assumption that 1) tweets are written in English and 2) even so, that tweets conform to the standard length of words in the English language.

This would imply that 132,052 words were tweeted in one second on Twitter. For comparison, *The Hunger Games* contains 99,750 words. J. K. Rowling used 76,944, 85,141, and 107,253 words in *Harry Potter and the Philosopher's Stone*, *Harry Potter and the Chamber of Secrets*, and *Harry Potter and the Prisoner of Azkaban*, respectively. Note, in one second more words were tweeted about Beyoncé's news than any one of these Harry Potter books! If this tweeting rate were kept for two seconds, then more words were tweeted than all three of these Harry Potter books combined.

Now, it's your turn. The Beyoncé tweet-fest was usurped in January 2012 during the NFL playoffs. In the Wildcard game, the Denver Broncos and Pittsburgh Steelers were locked in a tie, sending them to overtime. Only 11 seconds of extra time were needed for Tim Tebow to throw the longest overtime touchdown pass in what then became the

shortest overtime playoff game in the history of the NFL. The fervor in the Denver stadium shook the Rockies and created the Internet tsunami that followed. According to *USA Today* [20], thumbs were typing at a rate of 9,420 tweets per second on Sunday night as they announced the victory.

Challenge 3.3. *Making the same assumptions of average length of a tweet as we did for Beyoncé, how much textual content flew through Twitter from Tebow's inspirational overtime win? How does it compare to the 257,000 words in* Harry Potter and the Order of the Phoenix *or 119,000 words in* Twilight? *If this rate were maintained for 10 seconds, how would such a word count compare to the other books mentioned in this chapter?*

Returning the question that opened the chapter, would you take the million today or a penny? If you take the penny, you get to take home twice as much as the previous day over the next 30 days. One option will make you a multi-millionaire.

Half the Work

Keep in mind that rapid decay is also a powerful tool. To see this, let a friend choose a word in a book. After the word is selected, place about half of the book's pages in your left hand (or right if you prefer) and ask, "Is the word you selected in the pages that I hold in my left hand?" With the answer to this question, you immediately eliminate half the number of pages that could contain the desired word. Proceed accordingly, hold half of the non-eliminated pages in your hand and ask your next question. "Is the word you selected in the pages I now hold in my left hand?" Repeat this process until you have discovered the page that contains the word. Continue to ask questions which halve the number of possible words that have not been eliminated on the page until you have discovered the word. Remember to always state the question such that it can be answered by a yes or no response. With the power of exponential decay, the selected word soon emerges as the only possibility.

Will 30 questions be enough? In fact, we have allowed for far more than enough questions. By working backwards, this exercise shows a striking similarity to the amount of fortune Terentius carried with his trips to the emperor's treasury at the beginning of this chapter. If your friend chose a word from a book with only two words in it, only one question is necessary. Two questions could correctly identify the choice from among 4 words since the first question would reduce the possibilities to 2 words which requires one question to establish the chosen word. Continuing this, a document of 2^{30} words, which is over a billion, requires 30 questions. Now, *War and Peace* as translated by Constance Garnett contains just over 590,000 words. So, you are safe to pick any book you own and have a comfortable margin of error to find any word in it within 30 questions. Keep in mind that you generally won't be able to eliminate exactly half the remaining words. Practice, however, can increase the efficiency of the presentation and may even lead to an impressive party trick.

This technique is similar to a search algorithm in computer science called binary search, which can find an entry, called the search key, in a sorted list. Given such a list, take the first, last, and middle entries.

Do any of them equal the search key? If not, does the search key lie between the first and middle elements? If so, we now search the truncated sorted list that lies between the first and middle elements. Else, we search between the middle and last elements. We now have a new sorted list, find its middle element. Does it equal the search key? If not, truncate the list again and continue this process. Binary search can find the search key of a billion items by looking at 30 or less entries. In our technological world, searching data quickly can be an important and common practice. Think of the number of files on a personal computer. When we search our hard drive for a file with a particular name, fast search algorithms enable us to find the desired digital file quickly.

> **Challenge 3.4.** *Suppose the entire human population of the Earth (approximately 7 billion) were listed in a book of names and you were asked to find a single person's name and were allowed only to ask yes or no questions that would be answered honestly. How many guesses would be needed?*

4 . . . .

Infinite Detail

Let's start with a game.

1

2   3

Rules:
1. Place a dot halfway between square 1 and 2.
2. Roll a die and place a new dot halfway between your last dot and:
    - the middle of square 1 if you roll 1 or 2
    - the middle of square 2 if you roll 3 or 4
    - the middle of square 3 if you roll 5 or 6
3. Return to Step 2.

Play for a while! What shape emerges?

Mini-Me

The image that would emerge from the game that started this chapter, if you played the game on the previous page long enough, is the image in Figure 4.1 (a). Note, you'd have to play long and accurately enough to see this image emerge. Creating the image point by point can be difficult and, indeed, time consuming. This image is what is called a fractal and known as Sierpinski's triangle.

Let's look at an important feature of this triangle. Notice how the image contains 3 copies of the larger image. There is one at the top and two along the bottom. This can be more readily seen when each copy of the larger image is colored differently as seen in Figure 4.1 (b). Magnifying an object and seeing similarities to the whole is an important property of fractals. An object with self-similarity has the property of looking the same as or similar to itself under increasing magnification.

Broccoli exhibits properties of self-similarity. That is, smaller stalks look like the larger stalks of broccoli—albeit at a different scale. In what ways does a piece of broccoli exhibit this property? Will it exhibit such a property under any arbitrary amount of magnification? How about Serpinski's triangle? Does it exhibit properties of self-similarity? Will

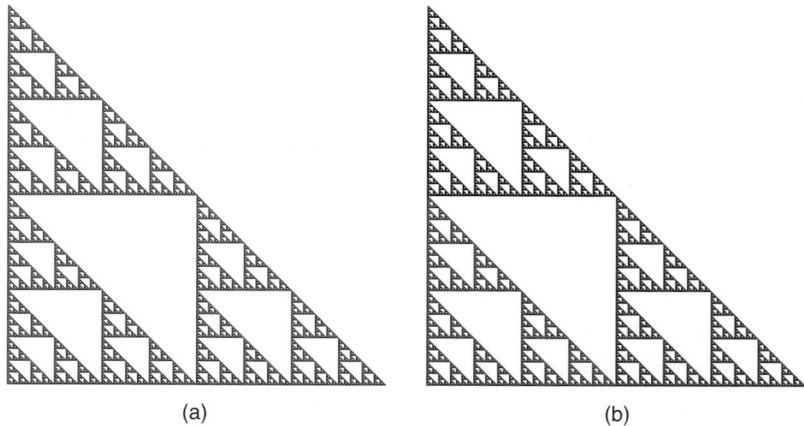

(a)   (b)

Figure 4.1. Sierpinski's triangle, a fractal, named after its founder, Waclaw Sierpinski. On the right, the image is colored so its property of self-similarity is more noticeable.

Figure 4.2. Broccoli supplies a real-life object with fractal-like attributes.

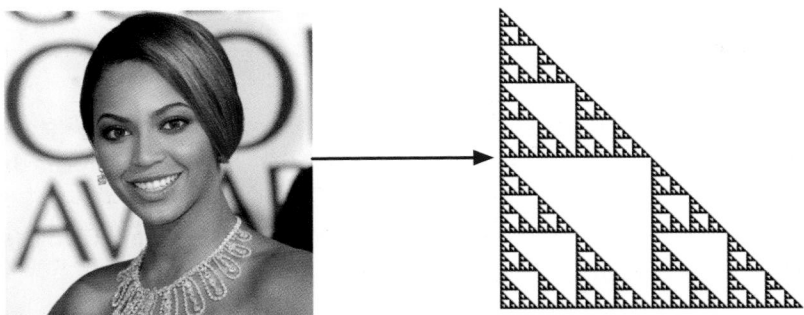

Figure 4.3. Can we start with a picture of Beyoncé and create Sierpinski's triangle?

it under any arbitrary amount of magnification? Sierpinski's triangle is created after an infinite number of loops of our previous game. As such, we could find copies of Sierpinski's triangle within the shape under any magnification.

Now, with a starting image and a photocopier or graphics program, let's create Sierpinski's triangle. We'll start with a picture of pop singer Beyoncé as seen in Figure 4.3.

To create the image, we create the following loop. To begin, let the current picture be the image of Beyoncé seen on the left in Figure 4.3.

1. Take your current picture and make 3 copies of the image reduced in size by 50%.

2. Construct a collage by placing the 3 images in the configuration seen in the table below:

Image 1	
Image 2	**Image 3**

3. Does your image look like Sierpinski's triangle? If so, stop. If not, loop back up to step 1 and think of your collage now as your current picture.

The image in Figure 4.4 is what we get after one loop of these steps. This doesn't look much like Sierpinski's triangle. So, we perform the loop again.

Now, we get the image on the left in Figure 4.5 and another pass through the loop produces the image on the right.

Figure 4.4. Creating a fractal with collage method.

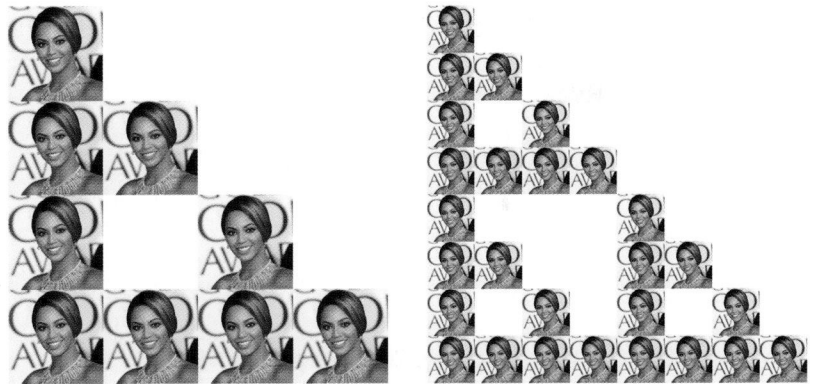

Figure 4.5. The next steps of making Beyoncé into Sierpinski's triangle.

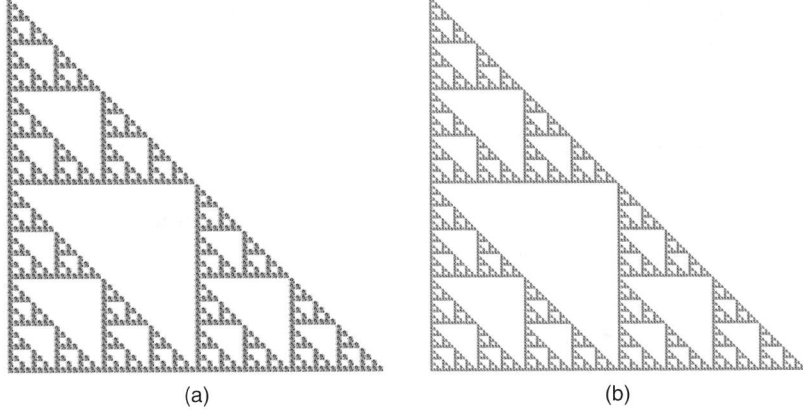

(a)   (b)

Figure 4.6. The next steps of creating making Beyoncé into Sierpinski's triangle.

It's a matter of taste as to when to stop. Two more loops through our steps produced the images in Figure 4.6. How'd we do turning Beyoncé into Sierpinski's triangle? Keep in mind that the first loop used 3 copies of the starting image. The second loop used $9 = 3^2$. The third loop used $3^3 = 27$. So, the image on the right in Figure 4.6 used $3^5 = 273$ copies of Beyoncé. After 10 iterations, we have 59,049 copies of the pop singer.

You will get the same results with any image. Select your favorite ones and give it a try.

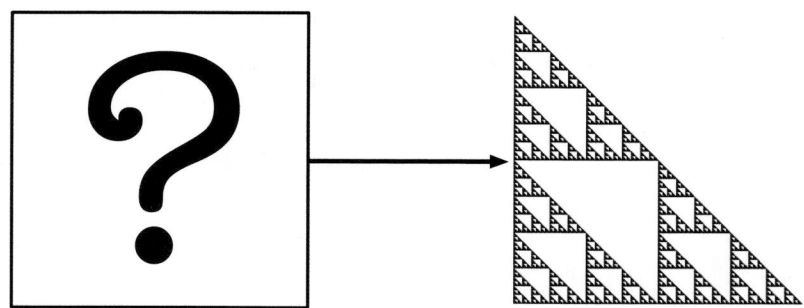

INFINITE BEAUTY

Fractals, like the one to the left, create beauty with very small amounts of storage due to their easy compression. Computer graphics was one of the earliest applications of fractals. In 1979, Loren Carpenter, of Boeing at the time, created the first computer movie of a flight over a fractal landscape. By 1982, Carpenter, now a senior scientist at Pixar, worked with a distinguished crew to create a fractal landscape of the Genesis planet in the movie *Star Trek II: The Wrath of Khan.*

Dicey Island

In this section, we extend the idea of fractals to create the landscape of a distant planet. First, let's create another fractal by performing the following steps with a pencil and paper:

1. Begin with a line segment that is 1 to 2 inches in length.
2. For each line segment in the current curve (which is initially one line segment):

   • divide the line segment into three segments of equal length,

- draw an outward pointing equilateral triangle that has the middle segment from the previous step as its base, and
- remove the line segment that is the base of the triangle from the step above.

3. Repeat step 2.

Let's perform the first iteration of these three steps together. Again, we begin with a straight line as seen below

Then, we divide the line segment into three segments of equal length, draw an outward pointing equilateral triangle that has the middle segment as its base, and remove the line segment that is the base of the triangle. This produces the shape below containing 4 line segments.

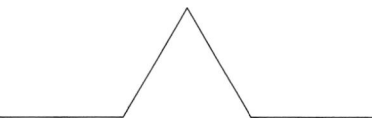

Continuing in this way produces the images in Figure 4.7 (a) and then (b). If we iterate infinitely many times, we produce the fractal called Koch's curve as seen in Figure 4.7 (c). An infinite number of steps isn't visually necessary since it becomes difficult for most (if not all) of us to produce very many steps by hand. So, we have a computer perform finitely many iterates and plot the resulting curve. Even with a computer, there comes a point when successive iterates no longer produce distinguishable changes in the resulting image. Such detail can only be seen through zooming into the image.

Let's explore this idea to create a coastline. Filling the upper and lower regions with blue and green creates the image in Figure 4.7 (d). There is a lot of structure. To create something more organic, let's add randomness.

We will start with a square located at $(0, 0)$, $(16, 0)$, $(16, 16)$ and $(0, 16)$ as seen in Figure 4.8.

Now, we follow these steps:

1. For each line segment (of which there are currently 4) compute the midpoint (x_m, y_m)

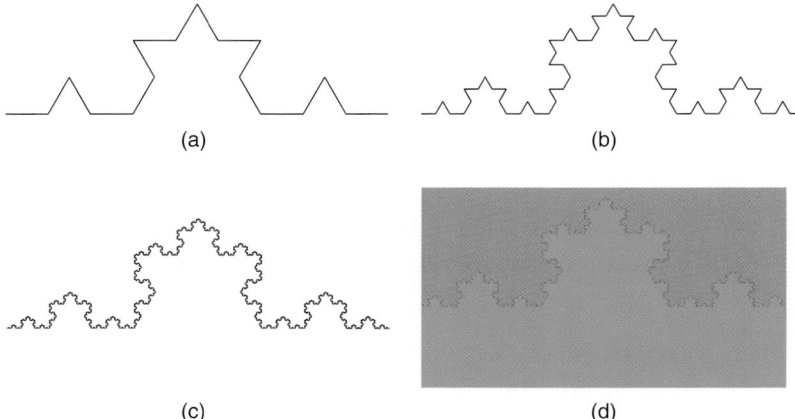

(a)

(b)

(c)

(d)

Figure 4.7. Various stages of creating Koch's curve along with a coloring of the image.

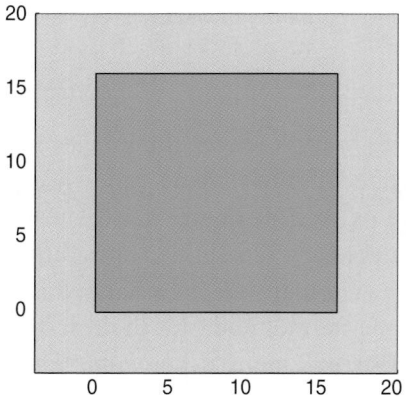

Figure 4.8. Starting with a square island.

2. Roll a die and if you roll

- 1-3, let $dx = 2$
- 4-6, let $dx = 4$

3. Roll the die again and if you roll

- 1-3, keep the dx above.
- 4-6, change dx to $(-dx)$.

4. Repeat steps 2 and 3 to find dy.
5. Your new midpoint will be $(x_m + dx, y_m + dy)$

For instance, I produce the following:

Endpoints	Midpoints	(dx, dy)	New endpoints
$(0, 0)$			$(0, 0)$
	$(8, 0)$	$(2, -2)$	$(10, -2)$
$(16, 0)$			$(16, 0)$
	$(16, 8)$	$(4, 4)$	$(20, 12)$
$(16, 16)$			$(16, 16)$
	$(8, 16)$	$(-2, 4)$	$(6, 20)$
$(0, 16)$			$(0, 16)$
	$(0, 8)$	$(2, 2)$	$(2, 10)$
$(0, 0)$			$(0, 0)$

Therefore, my square becomes the image in Figure 4.9 (a). Repeating the process but now for a roll of 1 to 3 in step 2 $dx = 1$ and a roll of 4 to 6 $dx = 2$, I got the image in Figure 4.9 (b).

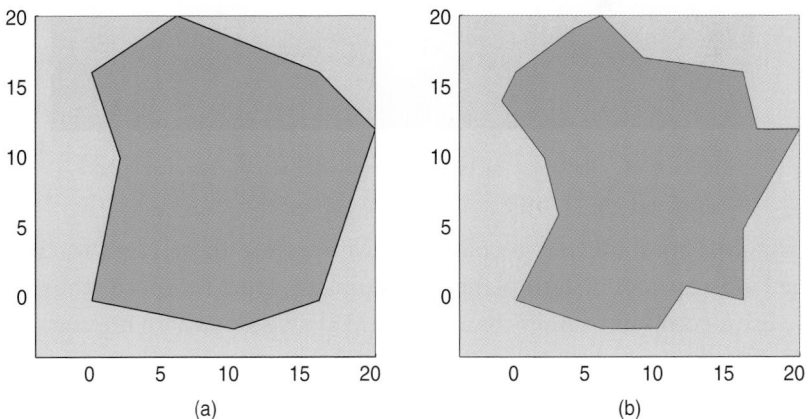

(a) (b)

Figure 4.9. Creating a fractal island.

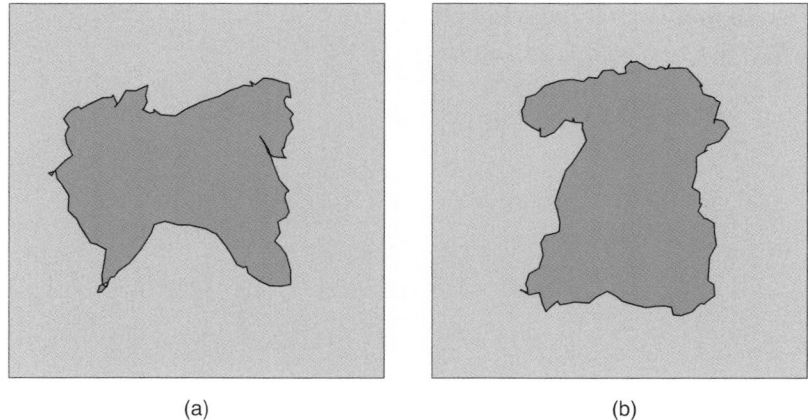

(a)   (b)

Figure 4.10. Creating a fractal island with a computer's random number generator rather than a die.

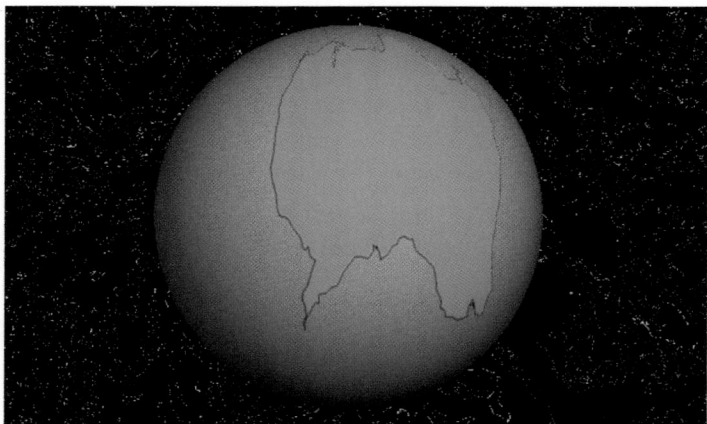

Figure 4.11. Placing a fractal island on a computer generated sphere.

Such shapes are fractal coastlines. Movies use these ideas to create the landscapes of distant planets. The images are a bit more realistic if we use a computer rather than a die and at stage k (where the square is stage 0), let dx and dy be a random number r chosen such that:

$$-\left(\frac{1}{2}\right)^k \left(\frac{7}{10}\right) \le r \le \left(\frac{1}{2}\right)^k \left(\frac{7}{10}\right).$$

(a)   (b)

Figure 4.12. Fractal images created by Ken Musgrave (a) and Bruce Clayton (b).

Two islands created with this approach are seen in Figure 4.10 (a) and (b).

Finally, I can take such an image and essentially wrap it around a sphere, as seen in Figure 4.11. This creates the fractal landscape for our distant planet. You can make your own landscapes with just a ruler, a die and some graph paper.

Fractal landscapes can also be extended to 3D, creating the topography of distant planets. To tour through a gallery produced by a leader in the field of fractal landscapes, search the internet for fractal images by Ken Musgrave. An example of his work appears in Figure 4.12 (a). Figure 4.12 (b) contains an image produced at a later time after fractal landscape technology had advanced.

5. . . . .

Plot the Course

IN THIS CHAPTER, we will see how functions like $y = 3x + 1$ and $y = 5 - x^2$ enable a computer to create a font like the one that comprises these words or plot the path of an Angry Bird through the air.

Mathematical Scribe

In days of old, calligraphers sat carefully lettering books with their artful creations. These masterpieces of script appear from cultures such as Indian, Tibetan, Persian, Islamic, Chinese, Japanese, and Western.

Let's learn the fundamental ideas behind creating fonts on a computer. Later, we will see a font that's a puzzle and could even be used as a secret message.

Fount of Fonts

Today's programs from PowerPoint to PhotoShop allow a user to choose a font to appear in various sizes from 6 pt to 24 pt to even 400 pt. Regardless of the size, the letters appear in perfect resolution. How is this done? The trick is that a font can be stored with only a few points

Figure 5.1. A scriptorium monk at work (left), from Lacroix, 1891. A decorated letter D (right) from an old bible.

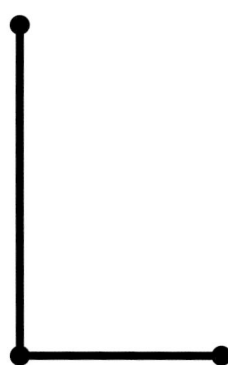

Figure 5.2. The letter L created with three points.

even though thousands of pixels will make up the complete curve. As an example, consider the letter L in Figure 5.2. If we know we connect the points with lines, then we need only 3 points regardless of how big the letter might be.

Let's create the letter "P" with straight lines and see how little information is needed. We need only store the coordinates of the endpoints of the lines as seen in Figure 5.3. If one draws with curves rather than straight lines, more points are needed for each curve, but the principle and low amount of storage is very similar. For more information, see [10].

To form the lines that comprise a letter, a computer only needs the set of points along with a listing of which points are connected.

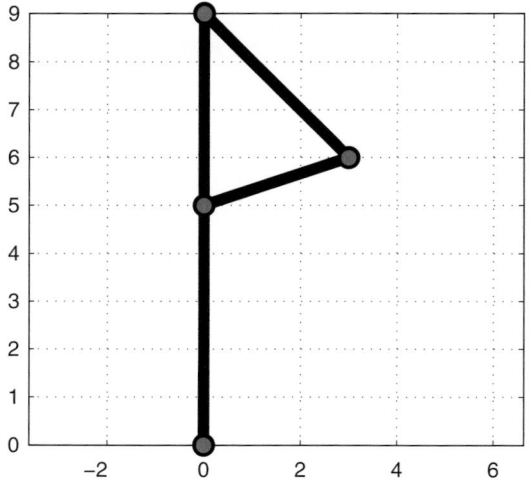

Figure 5.3. The letter "P" created with straight lines.

Efficient algorithms exist for drawing lines with computer pixels given two points. Still, let's find the equations of the lines in our letter. To do this, we'll need to find the slope m of a line between two points (x_1, y_1) and (x_2, y_2). This equals

$$m = \frac{y_2 - y_1}{x_2 - x_1}.$$

Once we compute m, we can find the equation of a line using this value and a point (x_1, y_1) on the line. The equation of the line is

$$y - y_1 = m(x - x_1).$$

Note that horizontal lines have a slope of 0 and vertical lines have an undefined slope.

Let's find the equations of the lines for the letter "P." There is the vertical line between $(0, 9)$ and $(0, 0)$. Since x equals 0 for both points, the equation of the line is $x = 0$. Next, the line between $(0, 9)$ and $(3, 6)$ has a slope equaling $m = (9-6)/(0-3) = -1$. Using point-slope form for the line, we find:

$$y - 9 = -1(x - 0) \text{ or } y = -x + 9.$$

In a similar way, the line between $(0, 5)$ and $(3, 6)$ has the equation $y = x/3 + 5$. So, this letter P is defined by the lines:

- $y = x/3 + 5$ from $x = 0$ to $x = 3$
- $x = 0$ from $y = 0$ to $y = 9$, and
- $y = -x + 9$ from $x = 0$ to $x = 3$.

Now, what if we want to double the size of the letter? Do we need any additional information? Note, we need only to multiply the coordinates we already have by 2. The points for the enlarged letter "P" are $(0, 0)$, $(0, 10)$, $(0, 18)$ and $(6, 12)$. So, we need only to store 4 points and know which are connected by lines to store this letter "P" for any size. No matter how big we make it, we can draw it in perfect resolution.

I've encoded a secret word by writing it only with the equations of the lines that define it:

- $y = 20 - \frac{10}{7}x$ from $x = 0$ to $x = 14$
- $x = 76$ from $y = 0$ to $y = 20$
- $y = 12$ from $x = \frac{194}{5}$ to $x = \frac{226}{5}$
- $x = 62$ from $y = 0$ to $y = 20$
- $x = 88$ from $y = 0$ to $y = 20$
- $y = \frac{5}{3}x - \frac{70}{3}$ from $x = 14$ to $x = 26$
- $x = 26$ from $y = 0$ to $y = 20$
- $x = 0$ from $y = 0$ to $y = 20$
- $y = 20$ from $x = 54$ to $x = 70$
- $y = \frac{5}{2}x - 85$ from $x = 34$ to $x = 42$
- $y = 12$ from $x = 76$ to $x = 88$
- $y = -\frac{5}{2}x + 125$ from $x = 42$ to $x = 50$

Challenge 5.1. *What is the secret word defined by the preceding line segments? You may want to be graphic and plot the word.*

Getting into Gear. Let's create a different font with points. This time equal-sized circular gears will be placed at the points as in Figure 5.4 (a). Now, we wrap a rubber band around the gears so it touches every gear once and is taut. What forms? Figure 5.4 (a) becomes the letter in (b).

How do I know you'd create an M from the gears in Figure 5.4 (a)? MIT computer science and mathematics professor Erik Demaine and his collaborators designed the gears so there is only one way to do it. [15] In fact, they created an entire alphabet in this way. With this in mind, let's look at the hidden message in Figure 5.5. Can you decode the word written in the gears? It may help to think of wrapping a rubber band around the gears.

> **Challenge 5.2.** *What word is encoded in the gears in Figure 5.5?*

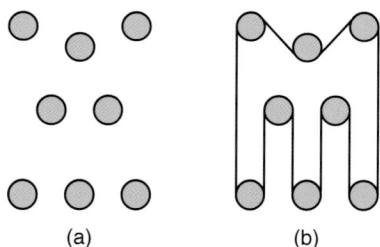

Figure 5.4. A conveyer belt font written only in gears (left) and with gears (right), making it easily legible.

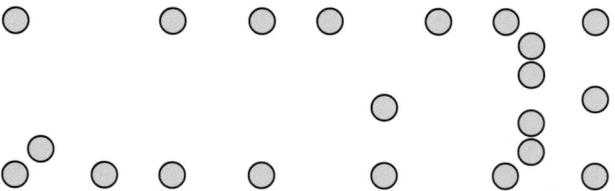

Figure 5.5. Can you decode the word written in the gears?

Throwing a Curve at Angry Birds

They're a phenomenon! They're addictive. Children and youth imprint their images on hats and t-shirts. Has a math craze begun? In a way no, as this refers to those airborne birds in the popular Angry Birds game. However, every launch of an incensed bird requires mathematical thinking. In fact, playing this video game can help teach fundamental ideas about quadratics, of course, if you are paying attention to the math.

On a smartphone or tablet, or through your browser, let's play Angry Birds. Each launch of a bird from the slingshot with the intent of a collision with a pig or fortified structure requires you to estimate a path of flight. You may not have computed a formula, but the bind is following the trajectory of one in the air.

Looking at the screenshot in Figure 5.6, how would you launch the bird so that it lands in front of the tower without hitting anything but the ground? Keep in mind that every time you change the launch angle for the bird, you are predicting its path. Now, launch the bird. Did you foresee the bounce? How close was the actual trajectory to your predicted one? Note the puffs of smoke that mark the path; they follow a nice smooth curve. Physicist Dr. Rhett Allain studied the game and verified in his article "The Physics of Angry Birds"

Figure 5.6. Screenshot of the smash hit Angry Birds.

(http://www.wired.com/wiredscience/2010/10/physics-of-angry-birds/) that our feathered friends travel without air resistance. So, a bird's path is a parabola.

Mathematically, parabolas can be written as $y = ax^2 + bx + c$. Choosing specific values for a, b, and c defines a curve. In Angry Birds, if you find the right a, b, and c, the birds enact their revenge on the pigs and their fortresses.

Can our knowledge of parabolas yield other information about this popular game? To begin, we must define a scale that can be consistent from screen to screen. We'll use a scale of 1 slingshot and adapt Dr. Allain's notation and write it as 1 AB as seen in Figure 5.7. Allain also found that 1 AB = 4.9 meters. So, the red bird is about 70 cm tall, which is one big, angry bird!

Parabolic motion, in the absence of air resistance, is defined by an initial position y_0, velocity v_0 and angle of launch θ, which forms the equation

$$y = y_0 + \tan(\theta)x - \frac{g}{2v_0^2 \cos^2 \theta}x^2$$

where g is called the gravitational constant and equals -2 AB/s^2. We need only to find v_0 to define an angry bird's path with this equation.

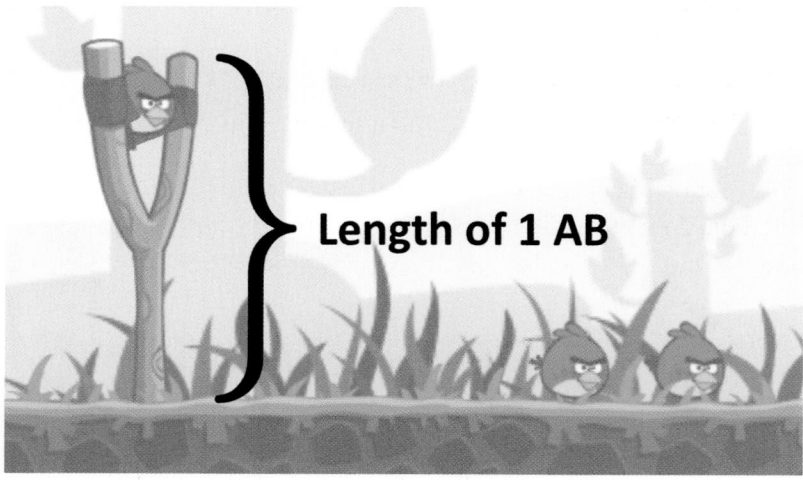

Figure 5.7. Defining the unit of 1 AB to equal the height of a slingshot in Angry Birds.

To do this, we use the pair of equations and introduce the variable for time, t,

$$x = v_0 t \cos(\theta)$$

$$y = y_0 + v_0 t \sin(\theta) - \tfrac{1}{2}g t^2 = 1 + v_0 t \sin(\theta) - t^2.$$

Note that the birds are launched one unit above the ground, implying $y_0 = 1$. Let's shoot a bird by pulling the sling straight back, parallel to the ground. Therefore, $\theta = 0$ and $y = 1 - t^2$. How long will it take the bird to reach the ground or $y = 0$ when $\theta = 0$? Simply solve $0 = 1 - t^2$ to find $t = 1$. When $t = 1$, $x = v_0$, for a bird shot at $\theta = 0$. So, we shoot a bird parallel to the ground and the distance that the bird lands from the slingshot (measured in AB units) equals v_0. When I shot a red bird several times, I found the average distance to be $x = 4.35$. Therefore, I computed a bird's initial velocity as 4.35 AB/s, which means the trajectory is described by

$$y = 1 + \tan(\theta)x - \frac{1}{(4.35)^2 \cos^2 \theta} x^2.$$

Choosing a launch angle of 45 degrees, a red bird will follow the path $y = 1 + x - 0.106x^2$.

We can use this equation to determine at what height a bird will be when it crosses the vertical line seen in Figure 5.8 if shot at 45 degrees. To answer this, we must determine where the dotted line crosses our x-axis. Remember, 1 unit equals the height of a slingshot. If we lay slingshots down one after another, we see that the dotted line is 5 units from the launching spot. So, we need only to set $x = 5$ in the formula $y = 1 + x - 0.106x^2 = 1 + 5 - 0106(25) = 3.35$. In Figure 5.9, we see the projected trajectories of a bird shot at 30, 45, and 60 degrees by the blue, red, and black lines, respectively.

What would be the highest point for a bird launched at 60 degrees, which corresponds to a bird following the graph of $y = 1 + 1.732x - 0.211x^2$? Such a point is called the vertex of the parabola. Such a point occurs at $x = -b/2a$. Since $a = -0.211$ and $b = 1.732$,

Figure 5.8. At what height will this angry bird cross the dotted line when shot at 45 degrees?

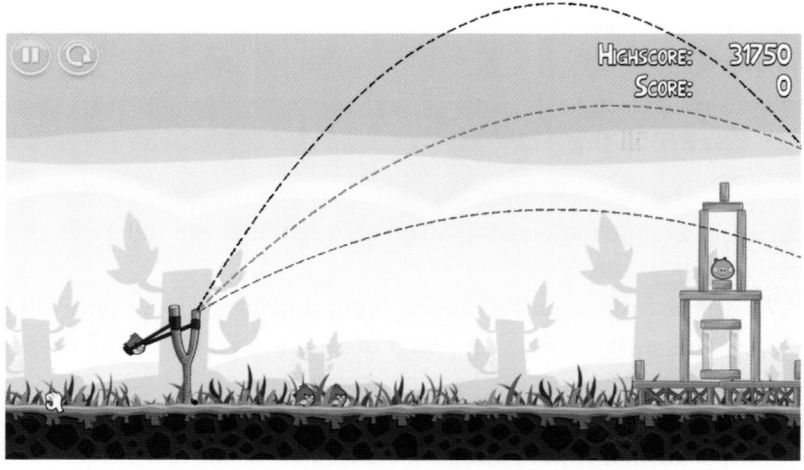

Figure 5.9. The path of an Angry Bird shot at 30, 45, and 60 degrees.

the vertex occurs at $x = 1.732/(0.411) = 4.214$ at a height equaling $1 + 1.732(4.214) - 0.211(4.214)^2 = 4.552$. Note from the picture above that the top of the initial screen is about 4 slingshots above the ground.

DIGITAL STUNT DOUBLES

Angry Birds uses a model of trajectory to model the path of the projectiles. The stunning special effects in movies utilize mathematical models, as well. For example, when a scene's action becomes too dangerous for a live actor or stunt double, a digital double is created on a computer to perform the tasks. To the left, we see a digital double of Obi-Wan Kenobi (played by Ewan McGregor) performing a stunt from *Star Wars: Episode II*. The hair on his head is simulated as strips of particles connected by springs.

Finally, we can find the point at which the bird will hit the ground when shot at 45 degrees. Note, this occurs when $y = -1$. So, we are interested in solving $-1 = 1 + 1.732x - 0.211x^2$. This occurs at $x = 9.234$. Will this be on the screen? You measure and answer the question.

Challenge 5.3. *Time for you to sling yourself into the calculations! If a bird is launched at 30 degrees, what is the highest point of the resulting trajectory?*

6. ·

Doodling into a Labyrinth

IN THIS CHAPTER, we begin with doodling as inspired by [8] and, by the end, use a math theorem to create a maze.

Doodling in Math Class

We begin with doodling. Place a pencil on a piece of paper and maybe even close your eyes. Create your doodle by drawing in big sweeping motions, keeping the pencil on the paper at all times. Remember that the longer you doodle the harder the exercise will be. On the left in Figure 6.1 is my doodle.

Let the math begin by placing a dot at every crossing point in your doodle. Also place a dot at both endpoints in your drawing, unless somehow you started and ended at the same spot. Count each dot and write that number on your doodle. In Figure 6.1 on the right, you can see the 11 dots for my doodle.

Next, count the enclosed areas of the doodle. Write this number on your paper. On the left in Figure 6.2 I found 9 in my doodle. Notice that the entire outside area of the piece of paper is counted as a space, which makes a total of 10. This will be important in a moment. Finally, count each segment between two dots. Write this number on your paper. My doodle has 19, which I label in Figure 6.2 on the right.

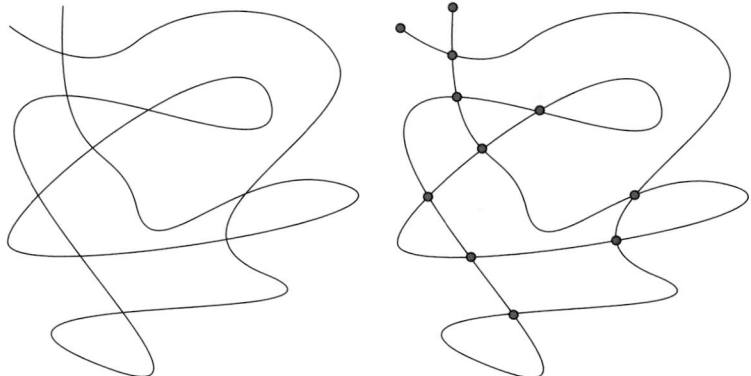

Figure 6.1. A doodle (left) and some math on it (right).

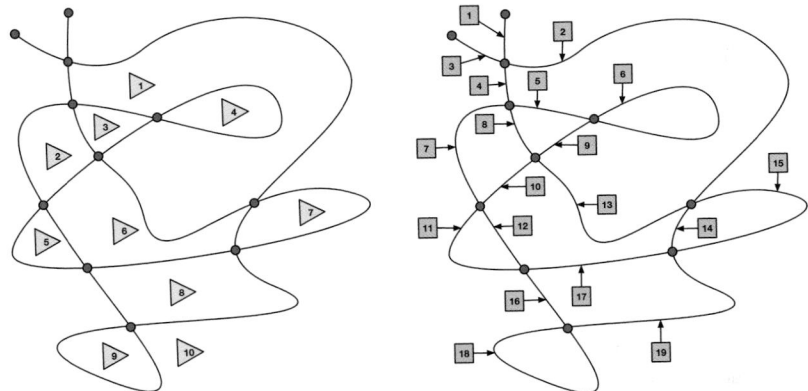

Figure 6.2. Counting enclosed areas and segments between dots in my doodle.

Add the number of dots and enclosed areas and subtract the number of segments. This gives you a number and you should have the same as me—the number 2. If you didn't, try counting again and check your addition and subtraction. Why would this happen? First, the dots are often called vertices, the enclosed areas are called faces, and the segments are called edges. If we denote these V, F, and E, respectively, then we will see the formula that Leonhard Euler proved

$$V + F - E = 2.$$

A MATHEMATICAL GIANT

Leonhard Euler was a famous mathematician who lived during the 1700s. He wrote hundreds and hundreds of pages on his mathematical discoveries. Among Euler's many works, he introduced the concept of a function and was the first to write $f(x)$ to denote the function f applied to the argument x. He also introduced the modern notation for e which is the base of the natural logarithm and the letter i to denote the imaginary unit.

This property is called the Euler Characteristic. Euler's proof shows this property will hold for infinitely many doodles. Such is the power and generality of mathematical theory. Try a bunch of doodles and you will see that you get the answer 2 every time.

A Touching Puzzle

Let's change gears and consider a puzzle that appears in [13]. The game is played by rearranging the order of the blocks numbered 0 through 9 seen in Figure 6.3 (a). Each block is assigned a score equaling the product of that block's value and the number of squares it touches with neighboring blocks. The score for a particular arrangement is the sum of the scores for all the blocks.

Let's compute the score of the initial arrangement in Figure 6.3 (a).

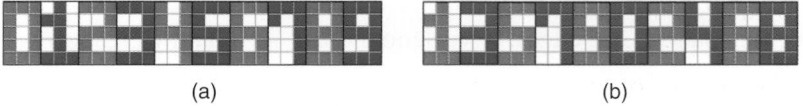

(a)   (b)

Figure 6.3. Two arrangement for the Touching Numbers Puzzle.

The block for the number zero touches 2 squares on the neighboring block. So the number zero block receives a score of $0(2) = 0$. The number one block touches 2 squares on its left and 1 square on its right, giving it a score of $1(2 + 1) = 3$. The number two block touches 4 squares of its neighboring blocks so its score is $2(4) = 8$. Continuing this for every number and summing up every block's score, we get

$$0(2) + 1(3) + 2(4) + 3(6) + 4(7) + 5(8) + 6(5) + 7(6) + 8(9)$$
$$+ \, 9(4) = 277,$$

which is the score of this initial arrangement.

Let's try another arrangement as given in Figure 6.3 (b). The number one block touches one square on its neighboring block; so it's score is $1(1) = 1$. The number three block touches 1 neighboring square on its left and 4 neighboring squares on its right. So, this block's score is $3(1 + 4) = 15$. The number five block's score is $5(4 + 1) = 25$. Continuing in this way leads to this arrangement's score, which equals

$$1(1) + 3(5) + 5(5) + 7(5) + 9(9) + 0(9) + 2(7) + 4(8)$$
$$+ \, 6(9) + 8(4) = 289.$$

We produced a higher score with this new arrangement, which leads to the challenge.

Challenge 6.1. *What arrangement produces the largest possible score? The smallest?*

This puzzle can be represented by a graph as seen in Figure 6.4, which, for simplicity, looks only at the blocks for the numbers 0, 1, and 2. We connect the number blocks with arrows or directed edges that have costs associated with them. The cost of an edge from i to j is the number of squares that touch if i is placed to the left of j. If you place the number one block to the left of the number two block, 1 square would touch. The corresponding edge in Figure 6.4 has a cost of 1.

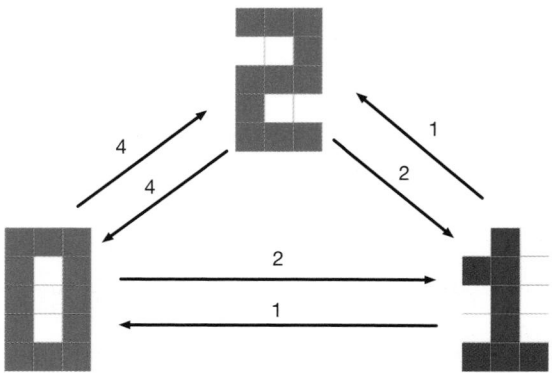

Figure 6.4. Graph of a small Touching Numbers Puzzle.

This formulation falls in the field of graph theory. Solving the puzzle corresponds to finding a path that visits every number block exactly once, where the sum of the weights of the traversed edges is as small as possible.

This puzzle, and in particular viewing it from this formulation, leads us to a game from the nineteenth century.

A Trip of a Puzzle

Let's play a game posed by Sir William Hamilton during the 1800s. For an octahedron as pictured in Figure 6.5 (a), find a path along the edges that visits every corner exactly once, except the origin node that is visited "first" and "last." After this, find such a path on the dodecahedron pictured in Figure 6.5 (b).

Sometimes, rather than visualizing such surfaces in 3D, it is easier to create what is known as a *Schlegel graph* of the solid. Figure 6.6 (a) depicts a Schlegel graph for a cube. Note how this corresponds to almost looking down into the cube from above.

We see Schelgel graphs of an octahedron and dodecahedron in Figure 6.6 (b) and (c), respectively. Our puzzle remains the same but can now be posed in 2D. We again search for a path along the edges of the graph that visits every vertex exactly once, except the origin node that is visited "first" and "last." Such a path is called a circuit.

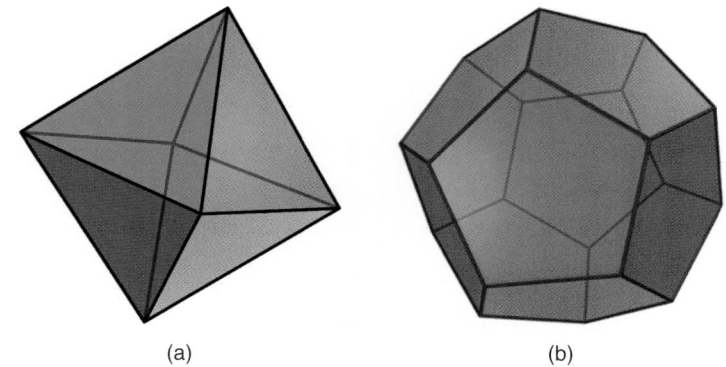

Figure 6.5. Octahedron and dodecahedron for a 3D puzzle.

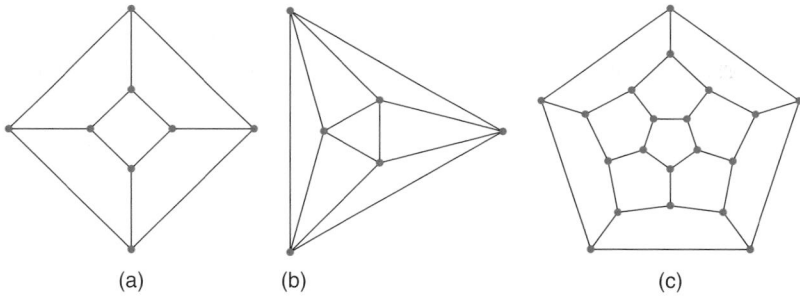

Figure 6.6. Schlegel graphs for a cube (a), an octahedron (b) and adodecahedron (c).

Sir Hamilton's puzzle, seen in Figure 6.7, was called an *Icosian Puzzle*. The pattern on the board is a Schlegel graph of a dodecahedron. The puzzle is the same that we just played.

If we put a weight on each edge and now search for a circuit with minimal sum of its edges, we have posed the Traveling Salesman Problem (TSP). It is called a TSP as the vertices of the graph can be regarded as cities and the weights as distances. The solution shows a salesperson the shortest route that visits all the cities and starts and ends at home.

Solving the TSP on even a moderately sized problem, such as 20 cities, is a problem that is well-known to be difficult. In 1962, Proctor and Gamble ran a contest that required solving a TSP on a specified

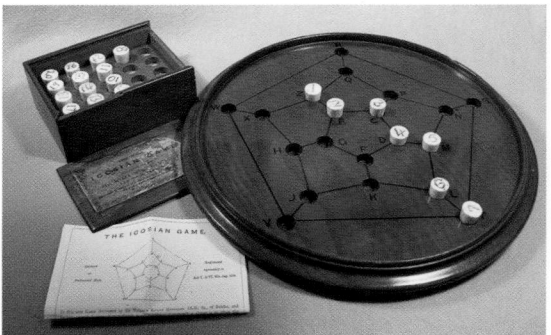

Figure 6.7. Icosian puzzle.

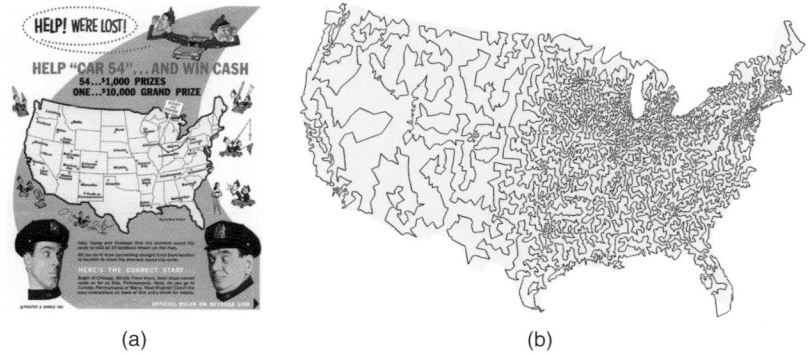

(a)   (b)

Figure 6.8. The TSP has an active history. Proctor and Gamble contest announcement (a) from 1962 and (b) contains the optimal tour found in 1998 of the 13,509 cities in the United States with populations greater than 500.

33 cities. An early TSP researcher, Professor Gerald Thompson of Carnegie Mellon University, was one of the winners. You can see the contest announcement in Figure 6.8 (a).

By 1987, the state of the art for finding TSP solutions had progressed considerably with Groetschel and Holland finding the optimal tour of 666 interesting places in the world. Just over 10 years later in 1998, Applegate, Bixby, Chvátal, and Cook found the optimal tour of the 13,509 cities in the United States with populations greater than 500. If you want to take such a tour, the route can be seen in Figure 6.8 (b).

(a)   (b)

Figure 6.9. A stippling (a) that serves as the cities for a TSP creating the route in (b).

For more information on solving TSPs and the history of tackling such an important and difficult problem, see [14].

We are now positioned to create some mathematical art.

A Maze of Math

Let's still create a route for a traveling salesperson but now the location of the cities is created from a stippling as seen in Figure 6.9 (a). If we start at a point and always follow a line segment from one point to another, what's the shortest path that visits all of the dots before returning to the starting point? The route is seen in Figure 6.9 (b). Such visual art, called TSP Art, was developed by Robert Bosch, Adrianne Herman and Craig Kaplan. [5, 6] In a moment we'll create mazes from such paths. First, try the maze in Figure 6.10.

Did you find the route from the start to the finish? If you did, you know something that I didn't when I made the maze. Still, I did know such a solution existed. How did I create this maze without knowing the solution? The Euler Characteristic from earlier in this chapter aided me.

The TSP Art image of Bart in Figure 6.11 (a) is constructed from line segments that form one continuous line. In other words, it could be created with one loop of string and nails placed as the endpoints of the line segments. The image has 5,000 vertices. There is one edge for every vertex so there are 5,000 edges, too. Recall the Euler Characteristic,

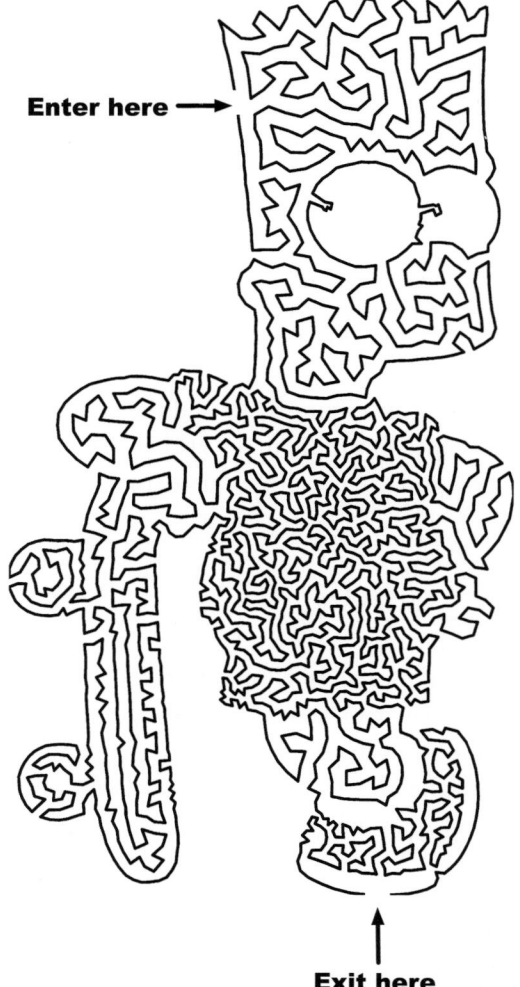

Enter here →

**↑
Exit here**

Figure 6.10. A maze of Bart.

which states

$$V - E + F = 2,$$

we know for Bart's image that:

$$5000 - 5000 + F = 2.$$

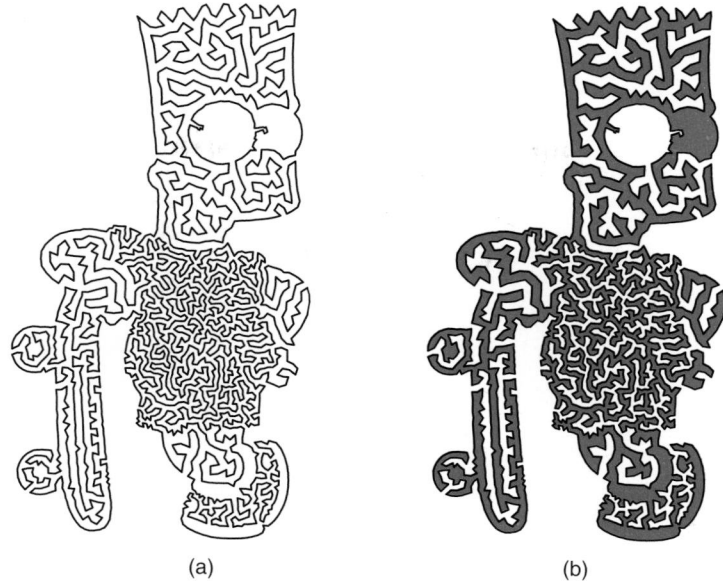

(a)   (b)

Figure 6.11. Bart Simpson in line art and half full.

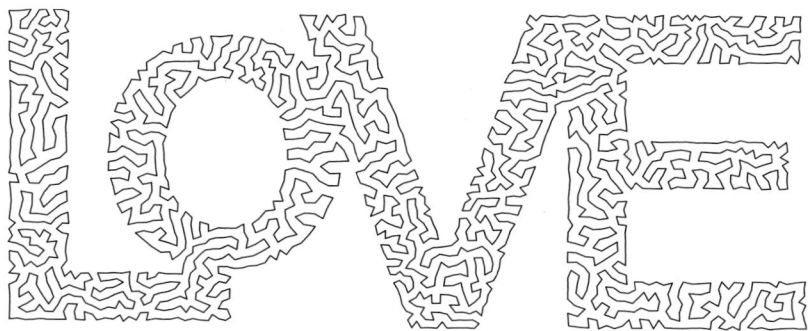

Figure 6.12. See the love!

Since $F = 2$, there are only 2 enclosed regions in this image. Look at Figure 6.11 (b) in which one region is filled with red.

So, how did I create the maze? I placed the start at some point along the boundary of the red region and the end somewhere else along the boundary of that region. As such, I know there is some path from the

Enter here →

Exit here ↑

Figure 6.13. TSP maze originating from a photograph.

start to the finish. What path? I don't know, but I have proven it exists. Though I created the maze and know the solution exists, the exact solution may continue to remain a mystery to me.

Challenge 6.2. *Find a start and end for a maze using the TSP Art in Figure 6.12. Remember, one region will be connected to the area outside the picture. The start and end should be placed along the boundary of the other region.*

Got time? Solve the maze in Figure 6.13, which originates from a photograph.

7 ....

Obama-cize Yourself

DURING THE 2008 PRESIDENTIAL ELECTION, a poster (see Figure 7.1) containing a stylized image of Barack Obama, designed by Los Angeles street artist Shepard Fairey, appeared with the words "hope" and "progress." It became synonymous with the campaign. In January 2009 as Obama became president, the Smithsonian Institution acquired Fairey's mixed media collage for the National Portrait Gallery.

Let's mathematically transform a digital image into a stylized portrait similar to Fairey's. We must first understand a storage scheme for color images. A picture is a collection of dots, called pixels. Each pixel is a combination of red, green and blue and can be represented by the triplet (r, g, b) where r, g, and b are values between 0 and 255. Keep in mind that 0 corresponds to the absence of the color and 255 represents full intensity. So, (255,0,0) is the fullest intensity of red, black is (0,0,0), (255,255,255) is white, and (80,0,80) is a dark shade of purple.

Turning Gray

As a subject of our art, I'll simply snap a picture of myself as I'm writing with the camera on my computer. I'll look off to the horizon knowing that Obama-fication is the end goal. See what you think of the snapshot

Figure 7.1. Shepard Fairey's image of Barack Obama from the 2008 presidential election.

in Figure 7.2 (a). You'll probably notice the grayscale image to the right of my picture. Can you anticipate why it is there?

A lot of colors combine to form my image, whereas the Obama portrait contains only four—particularly off-white, red, and two shades of blue. Fairey's collage groups large regions with a single color. The trick is determining which pixels to group. The darkest sections of Obama's portrait used the darkest hue of Fairey's four color palette. In fact, we really only need to measure a pixel's intensity rather than its actual color. A pixel colored (255,0,0) could be viewed as having the same intensity as a pixel visualizing (0,255,0).

So in the hope of my Obama-fication, let's convert my image to grayscale by computing the average of each pixel's red, green, and blue values. If a pixel is colored (173, 74, 236), then we replace this with the single value $(173+74+236)/3 = 161$. Grayscale contains only one color channel ranging from black (a value of 0) to white (a value of 255). You can see what my image would have looked like on a 1950s television in Figure 7.2 (b). Now, we need to collapse the grayscale coloring to only

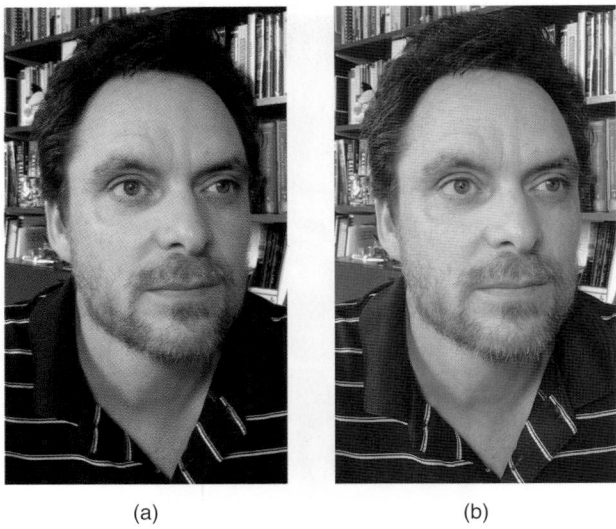

(a)   (b)

Figure 7.2. A snapshot for Obama-fication in color (a) and grayscale (b).

four intensities with the goal of re-coloring them off-white, red, and two shades of blue.

Two Sides of Obama-fying

We'll Obama-fy a picture in two ways. Both algorithms begin by sorting a picture's grayscale values into ascending order. Then, each element in the sorted list is assigned one of the four colors in the Obama image. Finally, the corresponding pixel in the original image is recolored with the assigned color from the sorted list. The methods differ in how such coloring is decided.

Method 1 - a balanced approach. This method uses all four colors in equal amounts. This is easily done by dividing the sorted list of pixel values into four equally sized groups, as seen in Figure 7.4. The pixels in group 1, 2, 3, and 4 are colored dark blue, red, light blue, and off-white, respectively. After this color conversion, my Obama-cized picture becomes the image seen in Figure 7.3 (b).

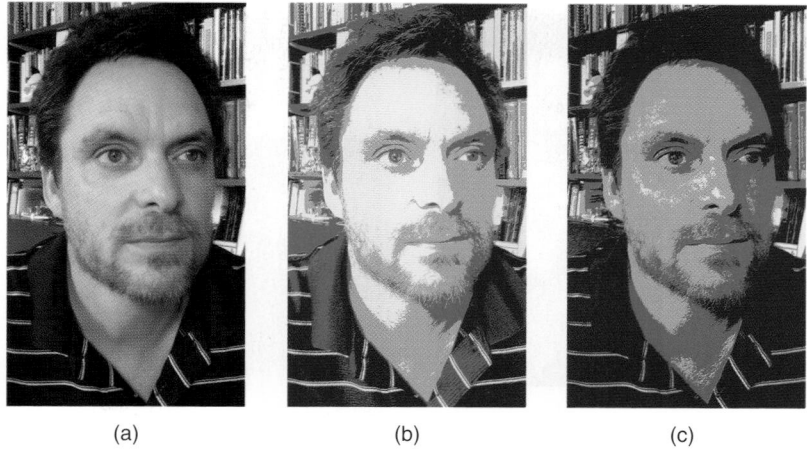

(a)   (b)   (c)

Figure 7.3. Image before and after Obama-fication using the colors equally (b) and by intensity (c).

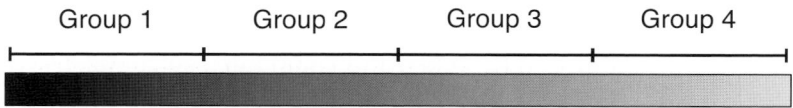

Figure 7.4. Sort and then group grayscale pixels to determine coloring for Obama-fication.

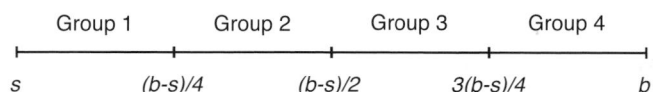

Figure 7.5. Sort and then group grayscale pixels in another way to determine coloring for Obama-fication.

Method 2 - an intense approach. This method colors by intensity and generally won't use the colors in equal amounts. After the sorting, find the smallest and largest grayscale values in the image and call them s (small) and b (big). Then, divide the interval between s and b into 4 equally space subintervals as seen in Figure 7.5. Therefore, every pixel in the image that has a value greater or equal to s and less than $(b - s)/4$ is in group 1. Note, we don't know how many pixels this

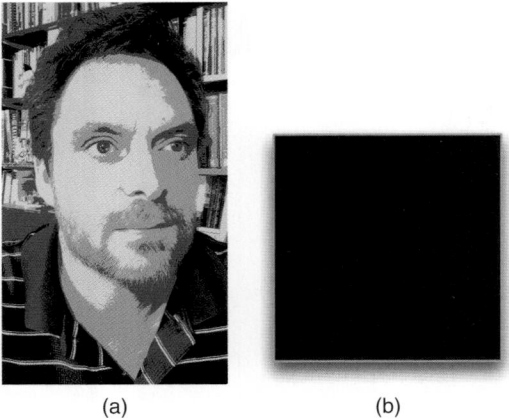

(a)   (b)

Figure 7.6. This image was produced the same division of pixels as Figure 7.3 (b) but with different coloring.

will be. The other groups are formed in a similar way. Then, as before, the pixels in group 1, 2, 3, and 4 are colored dark blue, red, light blue, and off-white, respectively. When the computer has completed this color conversion, my Obama-cized picture becomes the image seen in Figure 7.3 (c). Do you have a preference between Figure 7.3 (b) and (c)?

In this section, we created two simple algorithms to group an image's pixels. Keep in mind that if you like the Fairey image but prefer something less reminiscent of Obama, use other colors as in Figure 7.6 (a). Grouping data, often called clustering, is a popular and powerful field of computational mathematics and computer science. Clustering can group DNA samples to help diagnose illness, recognize communities within a large social network of people, or identify areas with greater incidences of a particular type of crime, which could allow better management of law enforcement resources. While Fairey used other means to create his memorable Obama portrait, these ideas allowed us to computationally create art by grouping data.

To see how different the two methods of this section can be, consider the black square in Figure 7.6 (b). How would methods 1 and 2 of this section Obama-fy this image?

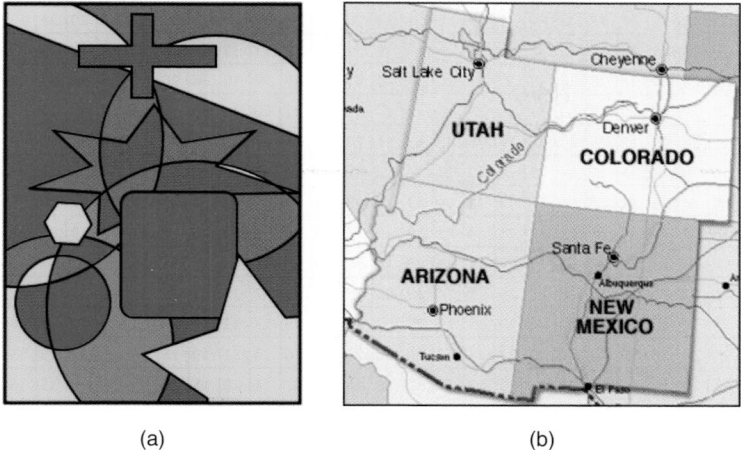

(a) (b)

Figure 7.7. Maps can be colored with four colors (a) although cartographers may choose more colors for their purposes.

Three and Four Color Palettes

Note, our Obama-fied images were colored with four colors. As a bit of an aside, we'll end this chapter mentioning the Four Color Theorem, which was the first major theorem proven using a computer. The proof is still not accepted by all mathematicians since it would be infeasible for a human to verify the proof by hand. The perceived lack of mathematical elegance by the general mathematical community was another factor, and to paraphrase comments of the time, "a good mathematical proof is like a poem—this is a telephone directory!" The Four Color Theorem states that any map can be colored with only four colors as we see in Figure 7.7 (a). Cartographers often use more than four colors for a variety of reasons as seen in Figure 7.7 (b).

Let's color an image with 4 colors. Rather than paint by number, we will paint squares in our table by interval. For example, a square will be colored black if the number it contains is between (or includes) 12 and 14. We'll have 4 colors in all. What image emerges? In the next chapter, we'll learn how to create such tables. Rather than numbers, we'll fill the squares with candies.

35	15	15	33	31	17	28	20	35	19	18	31	18	16	20	33	15	18	36	17	32	36	20	18	19	19
34	29	20	33	16	16	36	30	30	36	29	18	35	31	18	30	17	35	17	18	18	28	34	19	35	19
16	35	25	30	18	28	16	19	19	19	20	29	26	15	15	15	18	16	36	20	19	16	18	25	20	19
35	27	17	19	20	18	19	32	33	19	18	20	20	29	19	33	20	34	26	19	18	17	17	17	18	18
32	15	15	20	19	26	26	18	18	35	36	18	16	20	16	18	32	20	19	31	20	18	20	18	18	20
19	18	15	18	16	16	14	12	14	12	13	12	12	12	12	12	12	12	12	13	14	14	12	19	18	16
15	30	34	17	32	18	14	12	13	14	14	13	12	13	14	14	13	12	14	13	14	12	13	27	18	16
18	18	15	17	34	14	12	18	15	14	13	19	16	34	20	18	13	13	15	17	20	15	18	35	18	35
25	19	33	20	19	12	17	16	27	12	13	29	15	19	16	16	12	13	27	19	16	18	36	31	17	17
19	15	17	33	14	16	29	26	16	12	14	19	35	28	20	15	12	13	33	15	35	18	17	20	16	17
16	20	27	15	31	15	20	18	17	13	13	29	20	28	20	16	13	12	17	27	27	34	20	20	18	20
16	25	34	18	19	35	33	20	18	13	14	20	36	20	15	16	13	31	19	25	26	18	19	15	15	17
19	17	20	19	34	20	30	19	33	13	14	31	18	15	16	12	14	18	15	18	17	20	19	15	16	17
36	18	16	29	19	26	31	20	18	13	13	17	36	15	16	12	13	31	18	20	16	25	36	15	35	20
17	19	17	20	25	30	33	26	19	14	14	31	20	17	18	14	12	16	15	19	19	15	15	27	33	34
15	18	34	18	27	19	20	36	16	12	14	20	36	18	18	12	12	17	31	20	20	29	19	20	26	25
20	29	35	15	16	17	16	17	12	13	29	15	19	17	36	13	14	16	27	34	31	28	15	20	16	32
27	33	16	19	26	35	19	20	14	14	19	31	28	19	26	13	13	17	30	26	18	20	15	19	36	15
29	36	20	19	31	33	19	14	12	13	29	20	32	15	19	14	14	13	18	31	17	18	28	15	15	18
15	30	32	26	35	16	16	12	13	13	19	16	18	19	20	35	12	12	13	34	19	12	31	16	17	20
17	19	20	17	15	17	12	13	13	13	25	31	27	20	30	15	13	14	14	13	14	13	25	17	34	15
34	15	17	31	18	34	13	13	13	17	35	28	26	15	16	17	14	14	13	13	14	18	25	17	18	17
15	30	19	25	18	29	27	13	25	15	32	17	18	15	26	15	17	33	13	13	16	15	17	26	19	17
19	17	33	20	16	28	17	15	18	20	19	27	18	20	30	16	31	27	15	19	17	16	15	16	15	15
20	17	17	36	16	18	19	16	30	29	15	29	16	20	20	27	17	18	31	19	35	18	28	29	27	17
17	30	20	19	15	17	15	20	17	16	17	27	31	31	30	19	17	18	20	18	32	18	16	20	17	26

Above is a hidden image. Uncover it by coloring a square:

- black if the number in the square is between (or includes) 12 and 14,
- yellow if the number in the square is between (or includes) 15 and 17,
- orange if the number in the square is between (or includes) 18 and 20, and
- green if the number in the square is between (or includes) 25 and 36.

8.........................

Painting with M&Ms

Suppose you bake a sheet cake, slather the top with icing, and decide to ornament the top with M&Ms. In this chapter, we will learn how the sugary surface of your cake can be a workspace to perform some calculus or calculate an estimate to the value of π. Want a less math-centric cake decoration? By the end, we will learn how to create a portrait of friends, family or, as we will see, presidents with M&Ms.

Chocolate Calculus

Let's start with a simple chocolatey problem that will open a door to ideas of calculus. A Hershey's chocolate bar, as seen in Figure 1 (a), is 2.25 by 5.5 inches. We'll ignore the depth of the bar and consider only a 2D projection. So, the area of the bar equals the product of 2.25 and 5.5 which is 12.375 square inches.

Note that twelve smaller rectangles comprise a Hershey bar. Suppose you eat three of them as seen in Figure 8.1 (b). How much area remains? We could find the area of each small rectangle. The total height of the bar is 2.25 inches. So, one smaller rectangle has a height of $2.25/3 = 0.75$ inches. Similarly, a smaller rectangle has a width of $5.5/4 = 1.375$. Thus, a rectangular piece of the bar has an area of 1.03125, which

(a)   (b)

Figure 8.1. Math on a Hershey bar.

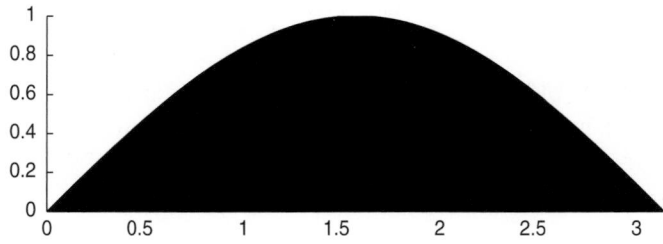

Figure 8.2. Graph of $y = \sin(x)$ for $x = 0$ to π.

enables us to calculate the remaining uneaten bar to have an area of $9(1.03125) = 9.28125$ square inches.

Let's try another approach. Remember that the total area of the bar is 12.375. Nine of the twelve rectangular pieces remain. Therefore, 9/12ths of the bar remains. I can find the remaining area simply by computing $9/12^*(12.375) = 9.28125$. Notice how much easier this is than the first method. Can we use an approach of this type in calculus?

A Riemann sum is a method for approximating the total area underneath a curve on a graph, otherwise known as an integral. The method was named after German mathematician Bernhard Riemann. The algorithm can be seen as nailing a row of planks side by side to cover the space under some curve. The thinner you make the planks, the closer they will be to covering only the area under the curve.

Let's change gears and use M&Ms rather than planks to find the area under the curve $y = \sin(x)$ over the interval from 0 to π. Color this region black as seen in Figure 8.2. The graph of $y = \sin(x)$ is contained entirely in the rectangle with corners at (0,0), (π, 0), (π, 1), and (0,1). This bounding rectangle has an area equal to π.

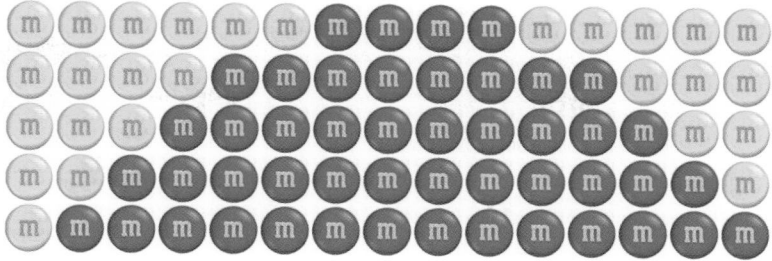

Figure 8.3. M&M mosaic for $y = \sin(x)$ over the interval $[0, \pi]$ with a 5 by 15 grid.

Now, place a grid of n by $3n$ non-overlapping squares over the graph. For each square, consider the ratio of black space to white space (that is, area under the curve to the area not under the curve). For now, we will measure this area, as best we can, with our eyes. If there appears to be more black than white in a grid box, place a green M&M in the box, else place a yellow M&M in the square. If you can't tell, flip a coin to choose a green or yellow M&M.

We'll use a 5 by 15 grid for our first approximation giving us 75 squares. Rather than visually approximating the area of each square, we'll have a computer take the ratio of black to white pixels in an image of the graph. If there are more black pixels, then a green M&M is placed. Using this technique places the green and yellow M&Ms in the configuration seen in Figure 8.3. With this candy mosaic, we can approximate A, the area under the curve, by counting the number of green M&Ms and the total number of M&Ms. That is,

$$A \approx \left(\frac{\text{no. of green M\&Ms}}{\text{total M\&Ms}} \right) * (\text{Area of bounding rectangle}). \quad (8.1)$$

The M&M mosaic in Figure 8.3 yields the approximation

$$A \approx \pi * (50/75) = 2.0944.$$

This works with the same reasoning that allowed us to find the area of the chocolate bar in Figure 8.1 (b). For the Hershey bar, we found that 3 of 12 or 1/4 of the rectangles were removed. In our M&M mosaic,

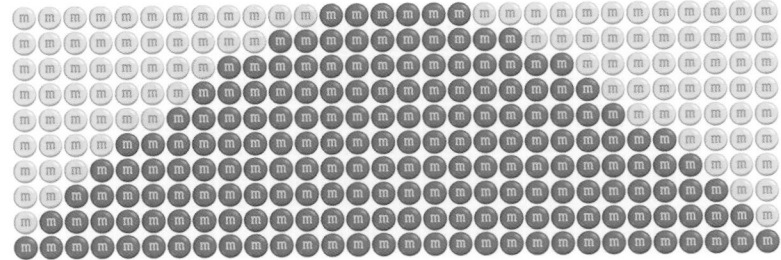

Figure 8.4. M&M mosaic for $y = \sin(x)$ over the interval $[0, \pi]$ with a 10 by 30 grid.

we find the fraction of squares that were filled with green M&Ms. As such, equation 8.1 computes the total area of the squares containing green M&Ms.

From calculus, the area under the curve is 2. Our first approximation was 2.0944. We will now improve our approximation.

To get a more accurate result, let's start over and use a larger grid. As such, the grid boxes will be smaller resulting in a more accurate approximation. For example, let's create a 10 by 30 grid, which results in the mosaic in Figure 8.4. Using equation 8.1, our new approximation for the area under the curve is

$$A \approx \pi * (194/300) = 2.0316.$$

Since 300 chocolate pieces increased accuracy, let's try more than a thousand with a 20 by 60 grid. This produces the mosaic in Figure 8.5 and an approximation of

$$A \approx \pi * (771/1200) = 2.0185.$$

Note that using 1,200 M&Ms as in Figure 8.5 requires having an image with at least 1,200 pixels, preferably many more. For a moment, let's assume we can easily produce such an image of any size. Suppose we were able to get our hands on one million delectable bites. Our approximation would be much closer to 2 than we've yet found. Now, imagine using one billion, one trillion or one septillion M&Ms. Would we reach 2? Remember that π is irrational and our fraction of green

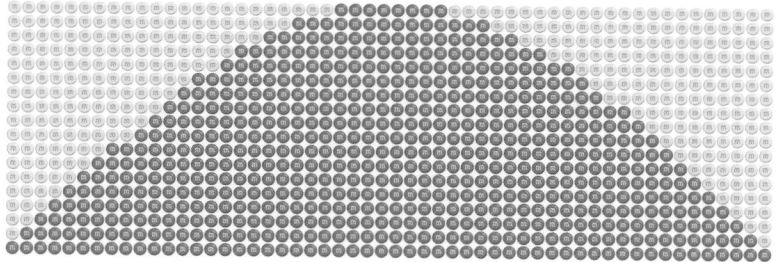

Figure 8.5. M&M mosaic for $y = \sin(x)$ over the interval $[0, \pi]$ with a 20 by 60 grid.

M&Ms to the total will always be rational. As such, the product of our irrational and rational numbers can never equal 2. However, if we could find approximations with more, more and more candies, we would, over time, continue to improve our approximation. Observing this behavior highlights the underlying principle of limits in calculus.

Our method works well but requires visually estimating area (which can be inaccurate) or counting pixels in an image (which is easiest for a computer). Let's alter our method so it is easier to do by hand and is not dependent on the size of the underlying image of the graph.

Again, we overlay the image with a grid. Now, only one point in each grid box will be used to determine what color M&M to place in a square. We'll take the top right corner of every grid box. Now, instead of comparing black and white pixels, if the chosen point in the grid box is under the curve, its associated grid box receives a green M&M. Else, the square is filled with a yellow M&M. Once the entire mosaic is created, the approximation is computed again using equation 8.1. For a 5×15 candy mosaic, we find

$$\int_0^\pi \sin(x) \approx 1.6755,$$

since 40 of the 75 M&Ms are green. Refining to a 10×30 grid yields an improved approximation equalling 1.8989.

With some clever manipulation we can significantly improve our mosaic method. The candy mosaics often contain a lot of squares containing some area above and also below the curve. Let's improve our

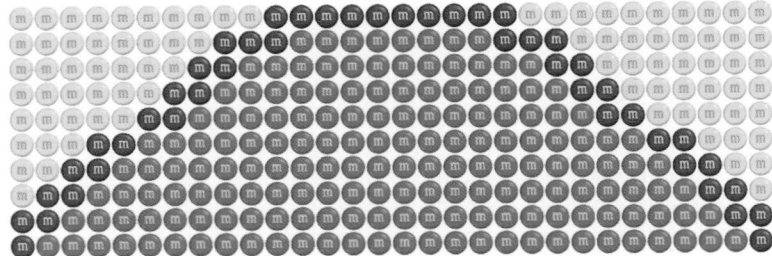

Figure 8.6. M&M mosaic for $y = \sin(x)$ over the interval $[0, \pi]$ using 3 colors.

approximation in these boxes. For all grid boxes that are not entirely above or below the curve, count half of the box as above the curve, and half as below. Our algorithm becomes the following:

- Any square that is contained entirely below the curve is filled with a green M&M.
- Any square that is contained entirely above the curve is filled with a yellow M&M.
- All other squares are filled with red M&Ms.

With the M&Ms placed, the approximation of the integral, A is given by

$$A \approx \left(\frac{\text{no. of green M\&Ms} + \frac{1}{2}(\text{no. of red M\&Ms})}{\text{total M\&Ms}} \right)$$

$$* (\text{Area of bounding rectangle}). \tag{8.2}$$

A candy mosaic constructed by this method for $y = \sin(x)$ is seen in Figure 8.6. This 10×30 grid yields the estimate 2.0211, improving our earlier 1.8989 estimate. Refining to a 20×60 grid yields 2.0106, which is again an improvement from the earlier method on the same grid.

Chocolate-Covered π

Let's use these high-calorie ideas to create an estimate to π. Let's take a quarter circle of unit radius, which has an area

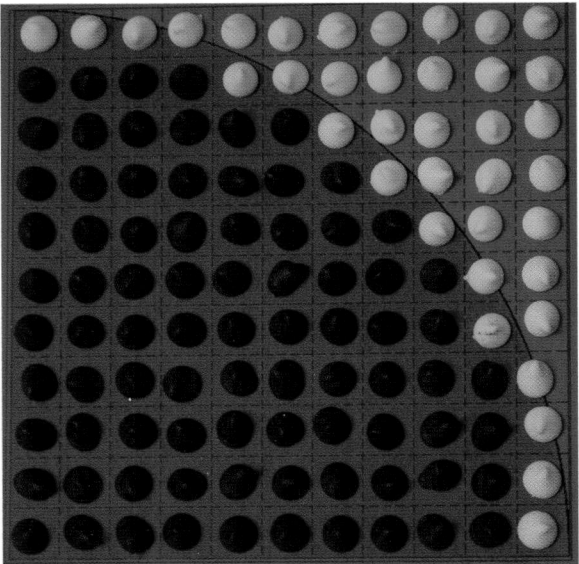

Figure 8.7. Estimating π on an 11 by 11 grid.

equaling $\pi/4$. This time, let's use milk and white chocolate chips. Overlay the graph with a grid of smaller squares. Place a milk chocolate chip on any square contained entirely in the circle. All other squares contain a white chocolate chip. For an 11 by 11 grid of squares, this produces the chocolatey mosaic in Figure 8.7. We are now ready to approximate π.

For the mosaic in Figure 8.7 (b), 83 of the 121 chocolate chips are milk chocolate. We estimate the area of the quarter circle as 83/121ths of the total area of the unit square. So, our approximation to $\pi/4$ is 83/121 also yielding $\pi \approx 4 \cdot (83/121) = 2.7438$.

Let's improve the calculation. We fill the grid boxes using the same procedure. Now, any square partially (but not fully) containing a portion of the circle contributes half a milk and half a white chocolate chip to the total number of chips used in the mosaic. All other squares contribute a full milk or white chocolate chip as before. How does this improve the calculation? Let's find out with the following challenge question.

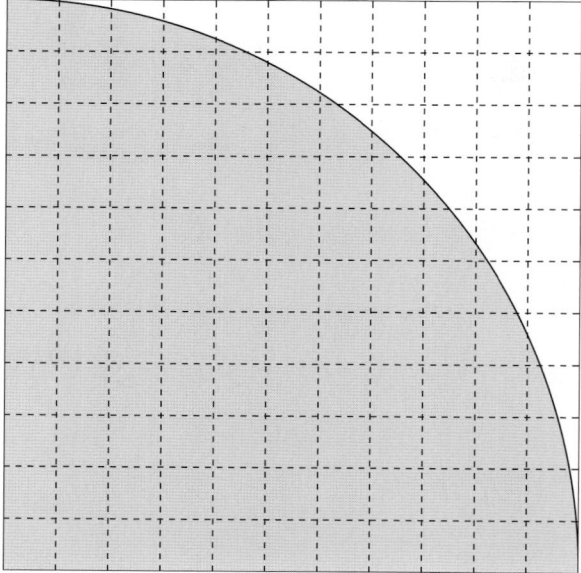

Figure 8.8. A grid for calculating π with chocolate chips.

Challenge 8.1. *Using the 11 by 11 grid in Figure 8.8, place a milk chocolate chip on any square contained entirely in the circle. All other squares contain a white chocolate chip. Any square partially (but not fully) intersecting a portion of the circle is counted as containing half a milk and half a white chocolate chip. Approximate π as the number of milk chocolate chips divided by the total chips used in the mosaic. How does this approximation to π compare to our earlier estimate of 2.7438 found earlier?*

Not a chocoholic? Skittles, Starbursts, Cheerios, or, for the calorie conscious, Sticky Notes could replace chocolate chips and be used to approximate π. In fact, you could even estimate this irrational number or compute an integral as you tile your bathroom floor or decorate a sheet cake.

Figure 8.9. Our palette of M&Ms for mathematical art.

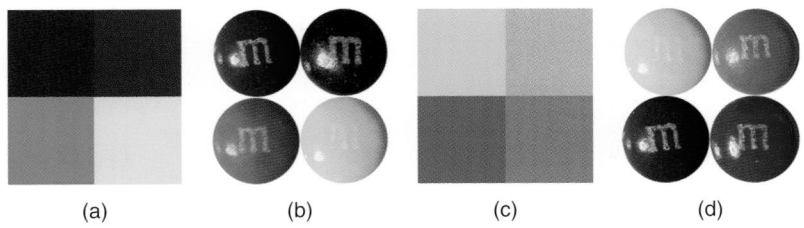

(a)	(b)	(c)	(d)

Figure 8.10. The regions in (a) and (c) are approximated by the M&M mosaics in (b) and (d).

Candy Mosaics

We just used M&Ms to estimate the value of an integral. Now, we will create mosaics with candy. While we have many choices of candy to use as tiles, we'll pick M&Ms since they melt in your mouth and not in your hand, giving us ample time to create our mosaic. So our palette of tiles is the group of M&Ms available in a standard bag of the treats as seen in Figure 8.9.

The image we want our tiling to approximate will be called the *target image*. Suppose we want to create an M&M mosaic to approximate the image in Figure 8.10 (a). Since each of the squares is the same color as an M&M, it is fairly clear that the mosaic in Figure 8.10 (b) is a good choice. Suppose instead we want to again use four M&Ms to create a mosaic but now the target image is Figure 8.10 (c). Visually, we can decide which candy pieces to use. How do we get a computer to choose on its own?

We again capture the three color intensities of a pixel as a coordinate triplet (r, g, b), where r is the amount of red, g is the amount of green and b is the amount of blue. The square in the upper left of Figure 8.10 (c) has the color $(208, 231, 32)$. We use the distance formula

TABLE 8.1.
The red, blue, green values of M&M candies used in our candy mosaics.

M&M color	red value	green value	blue value
blue	0	48	106
brown	77	33	8
green	50	170	83
orange	246	110	32
red	202	18	44
yellow	255	238	6

to find the M&M with the closest color. Using Table 8.1 as the colors of the M&Ms, the distance between the color of the upper left square in Figure 8.10 (c) and the color of a blue M&M is

$$\sqrt{(208 - 0)^2 + (231 - 48)^2 + (32 - 106)^2} \approx 286.$$

In contrast, the distance between (208,231,23) and (77,33,8), which the color of a brown M&M, is

$$\sqrt{(208 - 77)^2 + (231 - 33)^2 + (32 - 8)^2} \approx 238.$$

So, between these two M&Ms, brown is closer and a better choice. Not surprisingly, the yellow M&M has a distance of 54 and is the best choice. Repeating this for the other squares in Figure 8.10 (c) results in Figure 8.10 (d).

We can also replace a square of pixels with an M&M. For the square region, the average red intensity equals the average of the red values over the region. The average color of the region equals the average red, blue and green intensities over the region, computed in this way. This process allows us to replace a square of pixels by one color. We then approximate this square in our mosaic by the M&M with the closest color.

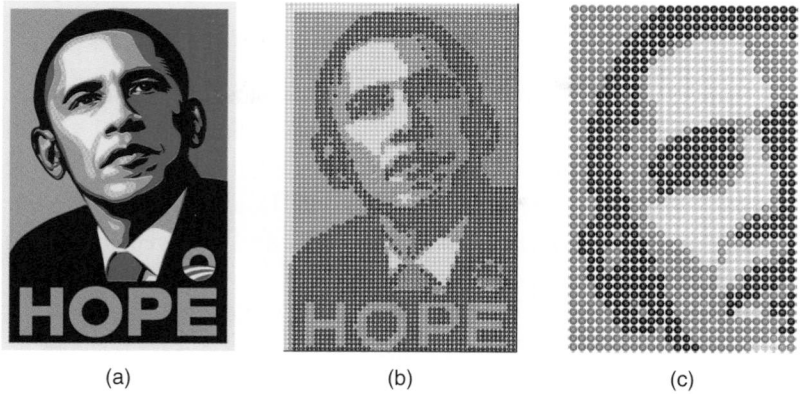

Figure 8.11. Shepard Fairey's image of Barack Obama (a) and an M&M mosaic of the poster (b) with a detail of the mosaic (c).

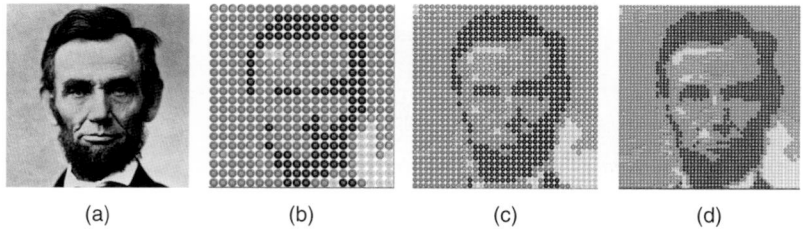

Figure 8.12. The target image of Abraham Lincoln (a) used to create the M&M mosaics with varying amounts of M&Ms in (b)-(d).

If we repeat this over the entire region, how does such an algorithm perform? See Figure 8.11 and decide for yourself with the image of President Barack Obama from Chapter 7. Then, let's switch to another president and create several mosaics of increasing refinement as seen in Figure 8.12.

Ready to try it on your own? Just keep in mind, the algorithm does not limit the number of times an M&M of a given color can be used. So, buy a big bag. You may even have some left over as a reward for your mathematical art. You also may wish to switch candies. Try Reese's pieces, Skittles, or even Froot Loops.

DOMINO MOSAICS

Dr. Robert Bosch of Oberlin College creates mosaics with dominos. He created the image of Marilyn Monroe below using 9 sets of dominos. Every domino in the sets is used in the image! None are left over. The picture of the Mona Lisa uses 96 sets. For more information on his work, see [4, 5] and his website http://www.dominoartwork.com/.

9. ·
Distorting Reality

COMPUTERS ALLOW FOR EASY METHODS OF image manipulation. From distorting an image to enhancing an image, one can change a picture with the click of a mouse. Let's start with a repeated image that becomes a bit of magic wrapped in a puzzle.

Vanishing Leprechauns and Pop-up Buildings

Counting to 15 is easy enough, unless you're counting leprechauns drawn by Pat Lyons. Below verify there are 15 of the Irish fairies.

If you cut out the puzzle along the straight black lines and interchange the left and right pieces on the top row, then you get the following configuration. How many leprechauns do you count now?

Look at the puzzle carefully and see if you can determine how a leprechaun suddenly disappeared on the page.

Rather than have a leprechaun vanish, let's have fourteen of the same object and then have a fifteenth appear. Count the fourteen letter Is below.

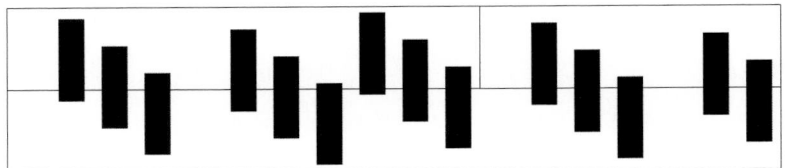

Swap places of the top two pieces in the puzzle and another letter is gained.

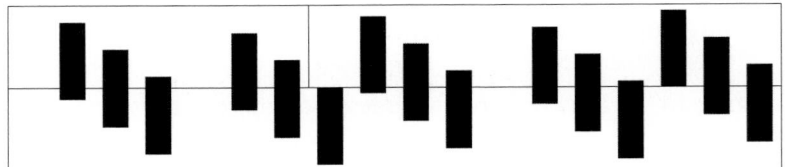

It can be difficult to spot what is happening in this picture. So, let's do some quick construction and have a building pop up.

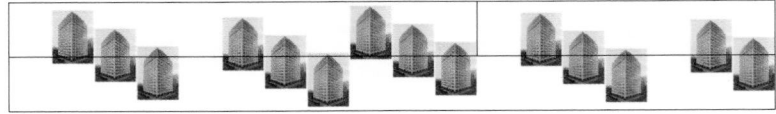

Swapping the pieces gives us another building. Look closely at the following image.

Can you spot any changes in the buildings in the first versus the second pictures?

How we pick up an additional image is more easily seen when we reorder the original set of images as seen below. First, we will take the letter Is.

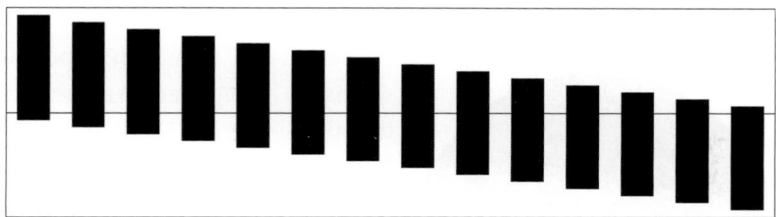

Swapping the pieces on the top row of the previous puzzle has the same effect as shifting the top piece as in the picture above as seen below. Simple enough, but notice how we pick up that additional image.

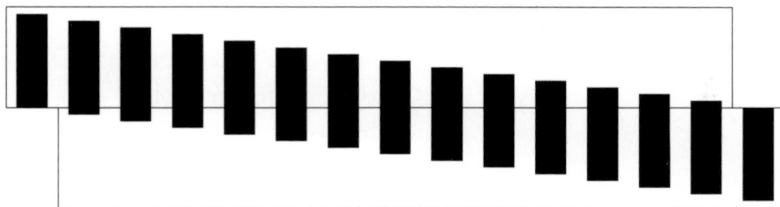

Let's look at this reordering again with the buildings.

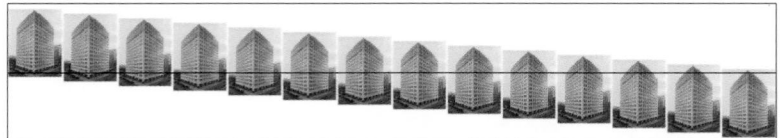

Notice how each image loses 1/14th of its total height. With this in mind, take a look the image of the shifted buildings.

MONKEY AROUND AT MOMATH

Another version of the type puzzle in this chapter places the images in a circle. A classic version is the "Get off the Earth" puzzle by Sam Loyd in which a person seems to disappear. At the National Museum of Mathematics in New York City, an exhibit, seen to the left, has two rings which can be placed in two positions relative to each other. In one position, there are twelve blue monkeys and thirteen red ones. In the other position, the numbers switch, so there are thirteen blue monkeys and twelve red monkeys. Did a monkey change color? The exhibit honors the work and writings of Martin Gardner who discussed such puzzles in [16].

With new insight on how the puzzle works, let's line up images of Kermit the Frog.

Let's swap upper pieces of the puzzle and see a Kermit the Frog appear.

Figure 9.1. An iconic photo of Marilyn Monroe

Did the images of the letter I, the building, or Kermit work the best? What image would you try? What if you used fewer images or more? Note you can always draw your own images, like the leprechauns, and make your own magic.

Math in a Photobooth

Let's perform a mathematical makeover on the iconic picture of Marilyn Monroe seen in Figure 9.1. We will compute such a transformation using polar coordinates, which, if you have an iPad or Mac, will be reminiscent of Apple's Photobooth program.

Often, we express the location of a point in two dimensions using Cartesian coordinates (x, y). Polar coordinates are simply another way to express such a location. For the point (x, y), we can express the same location in polar coordinates by calculating:

$$r = \sqrt{x^2 + y^2} \text{ and}$$
$$\theta = \tan^{-1}\left(\frac{y}{x}\right).$$

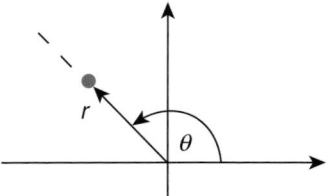

Figure 9.2. Expressing a point (x, y) in polar coordinates (r, θ) is simply an alternative way of locating a point in the plane.

As seen in Figure 9.2, r expresses the distance of the point from the origin and θ measures the counterclockwise angle of the point from the positive x-axis.

Have you ever wanted to grab someone's face and scrunch it up like a paper bag? Whatever your answer, it's easy in polar coordinates. First, we orient our photo such that the point $(0,0)$ is at the center of the picture. Using Cartesian coordinates for the moment, the upper righthand and lower lefthand corners of the picture are located at the points $(1,1)$ and $(-1, -1)$, respectively. Every pixel in our photo has an (x, y) location, which can also be expressed in polar coordinates as (r, θ). The picture will be transformed by moving each pixel located at (r, θ) to the point $(\sqrt{r}, \theta)$.

Let's try an example. Suppose we have a pixel located at the point $(x, y) = (0.25, 0.5)$. Let's convert this to polar coordinates. Recall $r = \sqrt{x^2 + y^2} = \sqrt{(0.25)^2 + (0.5)^2} = 0.5590$ and $\theta = \tan^{-1}(y/x) = \tan^{-1}(.5/.25) = 63.4$ degrees. In polar coordinates, our point is $(r, \theta) = (0.5590, 63.4°)$. Where does this point move in the plane? We find that the point $(0.5590, 63.4°)$ moves to the point $(\sqrt{0.5590}, 63.4°) = (0.7477, 63.4°)$. Both points are at the same angle but the new point is farther from the origin.

It may be helpful to see this visually. In Figure 9.3, we see 2 sets of points. The first set of points have polar coordinates $(0.4, \theta)$ where the θ values are equally spaced from 0 to 360 degrees; they are graphed with an open circle. Under our transformation, a point $(0.4, \theta)$ relocates to the point $(\sqrt{0.4}, \theta) = (0.6325, \theta)$; such relocated points are graphed with a solid circle. The second set of points have polar coordinates $(0.75, \theta)$ where the θ values are again equally spaced

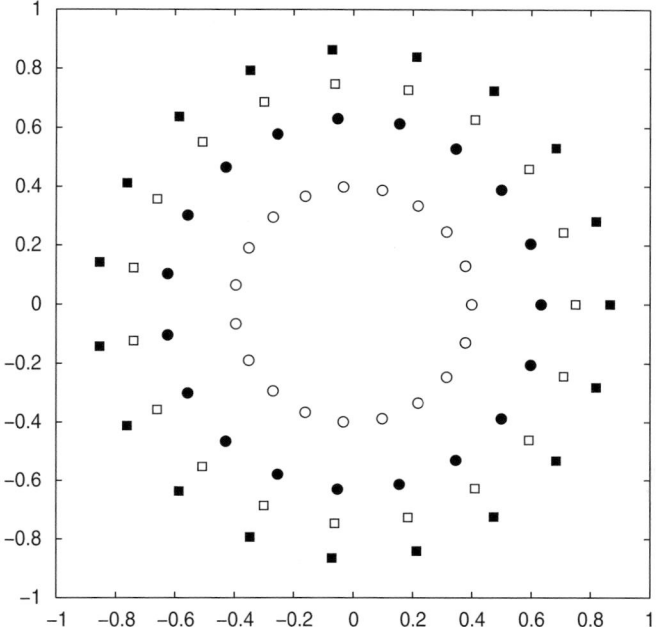

Figure 9.3. Points that are relocated via a polar transformation.

from 0 to 360 degrees. These points are marked with an open square. The transformation relocates a point $(0.75, \theta)$ to the point $(0.8660, \theta)$. These points are denoted with a solid square.

Now, let's stretch a face in another direction by relocating each pixel from its original position (r, θ) to (r^2, θ). It may be helpful to see this visually with the same 2 sets of points used in Figure 9.3. The points plotted with an open circle are relocated to the points marked with a solid circle. Similarly, the points marked with an open square relocate to the points plotted with a solid square. You can see the resulting graph in Figure 9.5.

Wondering about the effect on the image of Marilyn Monroe? Moving pixels from (r, θ) to $(\sqrt{r}, \theta)$ and (r^2, θ) results in Figures 9.4 and 9.6, respectively. Note that the picture was slightly translated in order to place Marilyn Monroe's nose at the origin.

Looking for another effect? Experiment more in polar coordinates and create your own extreme image makeover!

Figure 9.4. Warped image of Figure 9.1.

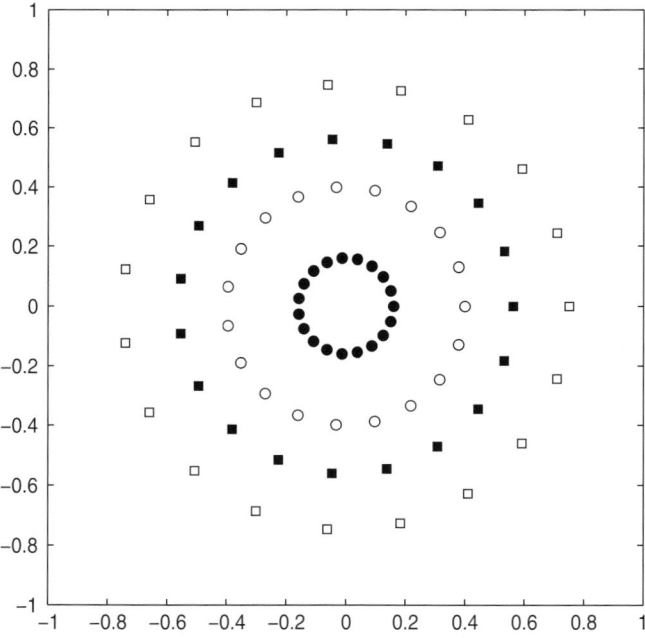

Figure 9.5. Points that are relocated via another polar transformation.

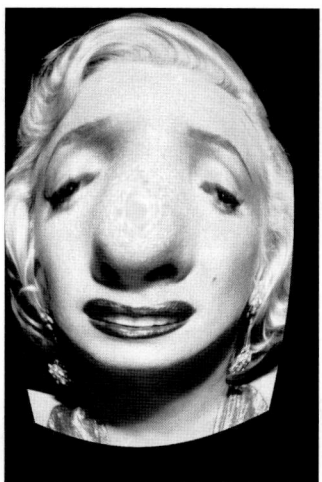

Figure 9.6. Warped image of Figure 9.1.

Coloring with Multiplication

Now, let's play with the color of an image. Would you like it black and white? yellowed, as if aged? possibly with more blue? or, as we will see, some other effect to capture interest. To begin, let's step from grayscale to color images.

The color information of images can be captured in matrices. Let's apply linear algebra to image manipulation. Suppose our image is n by m pixels. As in Chapter 7, we will store the color of each pixel in the image in a vector $\mathbf{v} = (r, g, b)$ where r, g and b represent the intensity of red, blue and green present in the corresponding pixel. A value of 0 equals the absence of that color and a value of 255 equals a full coloring of that value. So, (255, 255, 255) is a white pixel and (0, 0, 0) is black. Given this, we can save the color data of the entire image in a $n \times m \times 3$ matrix.

Let's begin by clowning around with the image in Figure 9.7 (a). To alter the image, we will take each pixel's color vector $\mathbf{v} = (r, g, b)$ and

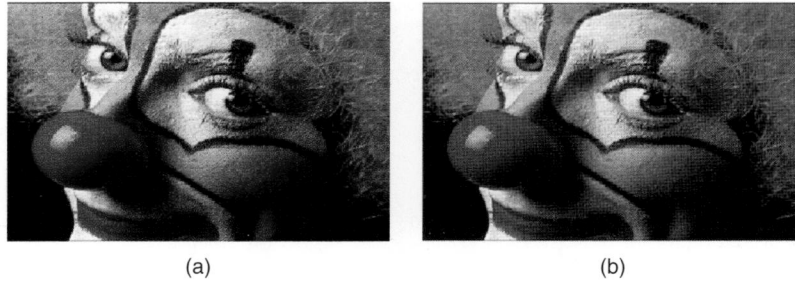

(a)   (b)

Figure 9.7. A color image of a clown (a) and a coloring of (a) after matrix multiplication (b).

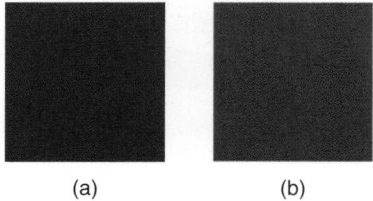

(a)   (b)

Figure 9.8. A color pixel and its value after applying a linear operator.

replace it with $B\mathbf{v}$, where

$$B = \begin{pmatrix} 1/3 & 1/3 & 1/3 \\ 1/3 & 1/3 & 1/3 \\ 1/3 & 1/3 & 1/3 \end{pmatrix}.$$

As an example, suppose a pixel has the following intensities $\mathbf{v} = (25, 77, 51)^T$. The pixel is seen in Figure 9.8 (a). We recolor the pixel by computing $B\mathbf{v} = (51, 51, 51)^T$, which corresponds to the image in Figure 9.8 (b).

To get more of a handle on the effect of applying B, let's try another matrix-vector multiplication. Notice $B(51, 204, 204)^T = (153, 153, 153)^T$. The matrix B averages the intensities of the red, green and blue values of the pixel and places this average in each element of the resulting vector. Visually, this is a method, although not necessarily the best, of creating a grayscale of an image. What happens if we apply B to the color data that created the image in Figure 9.7 (b)? You may have already guessed—it creates the grayscale version of the image in

 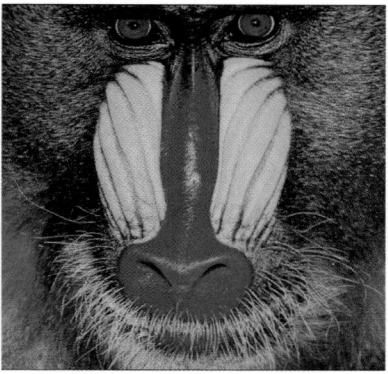

Figure 9.9. A color image of a mandrill (a) and the image after a linear transformation is applied.

Figure 9.7 (a). Sometimes, B is said to *desaturate* the image as it essentially squeezes the color information from the image.

A Warhol Transformation. Now, we will work with an image of a mandrill as seen in Figure 9.9. Let us apply the matrix (which operates as a linear transformation) to the color information for this image.

$$T = \begin{pmatrix} 0 & 1 & 0 \\ 0 & 0 & 1 \\ 1 & 0 & 0 \end{pmatrix}.$$

Again, let us look at one pixel's color data contained in a vector $\mathbf{v} = (25, 77, 51)^T$. Then the resulting pixel will have intensities $T\mathbf{v} = (77, 51, 25)^T$. We can analyze the effect of T on an arbitrary pixel by multiplying T by the vector $(r, g, b)^T$ which equals $(g, b, r)^T$. Can you visualize the change? You can see it in Figure 9.9 (b).

Let's create another matrix S which when multiplied by $(r \ g \ b)^T$ equals $(r \ g \ 0)^T$. The corresponding matrix is

$$S = \begin{pmatrix} 1 & 0 & 0 \\ 0 & 1 & 0 \\ 0 & 0 & 0 \end{pmatrix}.$$

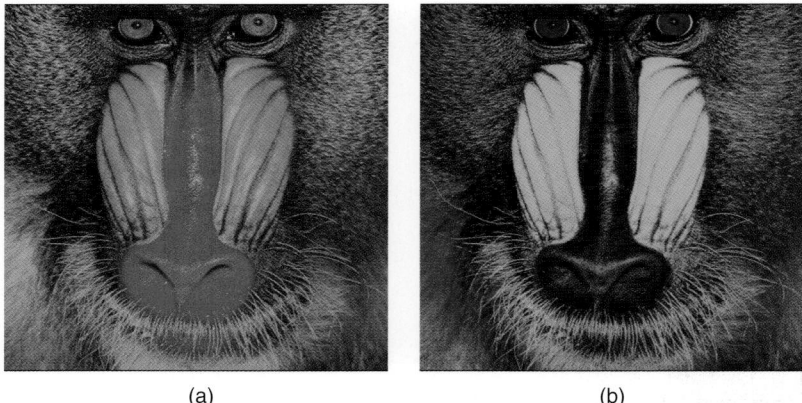

(a) (b)

Figure 9.10. Images created via linear transformations of the color data of the mandrill image in Figure 9.9 (a).

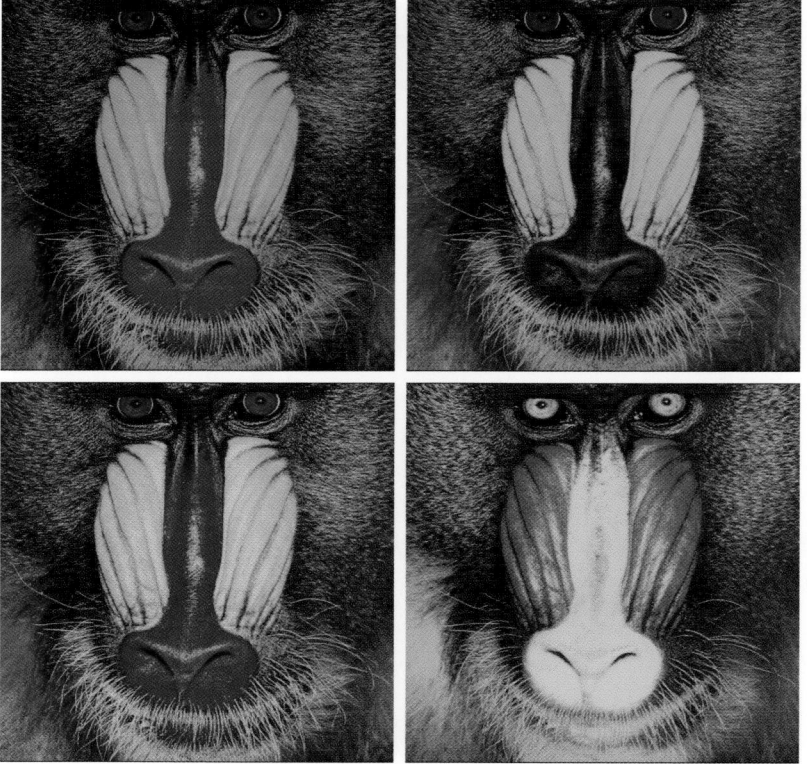

Figure 9.11. Images created via linear transformations of the color data of the mandrill image in Figure 9.9 (a).

Applying S removes the blue intensities from every pixel in the image. Applying S to the mandrill image results in the image in Figure 9.10 (a). For fun, let's form the matrix by multiplying S and T, $C = S(T)$. The resulting image of the mandrill appears in Figure 9.10 (b).

Armed with linear transformations, we can now play with the images until we acquire pictures of our liking. In fact, we can even create a collage of pictures reminiscent of the artwork of Andy Warhol. For such an image produced mathematically, see Figure 9.11.

Challenge 9.1. *Suppose a pixel's color intensities are stored in a vector* **v**. *We change the coloring by performing* **w** $= B$**v**. *Could we create another matrix* U *that could undo this coloring by* U**w**? *We want this product to equal our original* **v**. *Can this be done?*

10.......................

A Pretty Mathematical Face

HOW WOULD YOU DESCRIBE YOUR facial appearance? Granted, it may depend how long ago you awoke and if you have had your morning shower or coffee. Still, would you consider yourself a cross between Tom Cruise, Audrey Hepburn, and Clark Gable, or Robert Downey Jr. and Ellen DeGeneres? Let's see how mathematics can help answer such questions.

We'll work with a library of grayscale images of the 16 famous people seen in Figure 10.1. Our goal will be to find the combination of these pictures that best approximates a target image.

Recall from earlier in the book that the grayscale information of the image in Figure 10.2 can be stored in the following matrix:

$$A = \begin{pmatrix} 30 & 87 \\ 128 & 150 \\ 246 & 57 \end{pmatrix}.$$

Recall also that our grayscale values are all integers from 0 to 255 where 0 is black and 255 is white.

In this chapter, we will store the grayscale information in a vector. So the image information for Figure 10.2 would be stored in

Figure 10.1. Library of famous faces used in facial recognition.

Figure 10.2. Visualizing the grayscale data contained in a matrix.

a vector:

$$\mathbf{v} = \begin{pmatrix} 30 \\ 87 \\ 128 \\ 150 \\ 246 \\ 57 \end{pmatrix}.$$

Given that our 16 library images and 1 target image are vectors, we want to find the combination of the library images that satisfies the linear system:

$$x_1\mathbf{c}_1 + x_2\mathbf{c}_2 + \cdots + x_{16}\mathbf{c}_{16} = \mathbf{t}. \tag{10.1}$$

where $\mathbf{c}_i$ and $\mathbf{t}$ are the vectors storing image information for the ith library photo and the target image, respectively. The numbers

$x_1, x_2, \ldots x_{16}$ give us the amount of the corresponding image to add into the sum.

Let's look at a small example. Suppose $x_1 = 0.25$ and $x_2 = 0.61$, and our vectors are

$$\mathbf{v}_1 = \begin{pmatrix} 203 \\ 9 \\ 246 \\ 217 \\ 168 \\ 239 \end{pmatrix} \text{ and } \mathbf{v}_2 = \begin{pmatrix} 245 \\ 36 \\ 124 \\ 108 \\ 205 \\ 234 \end{pmatrix}.$$

The vectors $\mathbf{v}_1$ and $\mathbf{v}_2$ correspond to the images in Figure 10.3 (a) and (b). We can now form our target image,

$$\mathbf{t} = 0.25\mathbf{v}_1 + 0.61\mathbf{v}_2 = 0.25 \begin{pmatrix} 203 \\ 9 \\ 246 \\ 217 \\ 168 \\ 239 \end{pmatrix} + 0.61 \begin{pmatrix} 245 \\ 36 \\ 124 \\ 108 \\ 205 \\ 234 \end{pmatrix} = \begin{pmatrix} 200.2 \\ 24.2 \\ 137.1 \\ 120.1 \\ 167.1 \\ 202.5 \end{pmatrix},$$

which corresponds to the image in Figure 10.3 (c).

In equation 10.1, the variables $x_1, x_2, \ldots, x_{16}$ are the unknowns. Let's express this system of equations in another form by creating the matrix C where the first column of C equals $\mathbf{c}_1$, the second column of C equals $\mathbf{c}_2$ and so forth. We will solve the same problem as equation 10.1

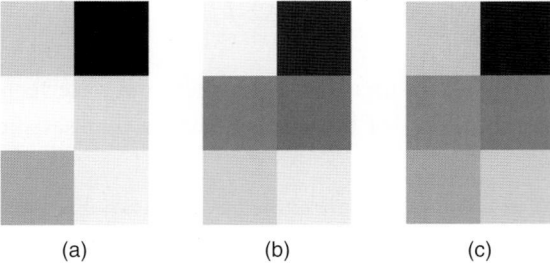

(a)   (b)   (c)

Figure 10.3. The grayscale image in (c) is a weighted sum of the image information in (a) and (b).

if we solve the linear system:

$$C\mathbf{x} = \mathbf{t}, \text{ where } \mathbf{x} = \begin{pmatrix} x_1 \\ x_2 \\ \vdots \\ x_{16} \end{pmatrix}. \tag{10.2}$$

We can see this from our earlier example since

$$\begin{pmatrix} 203 & 245 \\ 9 & 36 \\ 246 & 124 \\ 217 & 108 \\ 168 & 205 \\ 239 & 234 \end{pmatrix} \begin{pmatrix} 0.25 \\ 0.61 \end{pmatrix} = \begin{pmatrix} 200.2 \\ 24.2 \\ 137.1 \\ 120.1 \\ 167.1 \\ 202.5 \end{pmatrix}$$

ORIGINS OF MATRIX METHODS

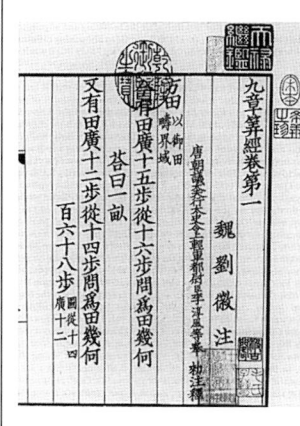

The *Jiuzhang Suanshu* is a Chinese manuscript dating from approximately 200 BC and containing 246 problems intended to illustrate methods of solution for everyday problems in areas such as engineering, surveying, and trade. (To the left, we see the opening of Chapter 1.) Chapter 8 of this ancient document details the first known example of matrix methods with a method known as *fangcheng*, which is what would become known centuries later as Gaussian elimination. [21, 2]

Chances are that there does not exist a combination of the library of images that can exactly replicate the target image. So, we cannot attain equality in the linear system in equation 10.2. Instead, we'll solve

$$C\mathbf{x} \approx \mathbf{t}.$$

What is meant by that approximation symbol? In terms of our library images, we want to find the sum that "best" approximates the target image. Mathematically, we will choose $\mathbf{x}$ by looking at the vector $\mathbf{r} = \mathbf{t} - C\mathbf{x}$. The first thing to notice about $\mathbf{r}$ is that it will equal $\mathbf{0}$ if there exists a combination of the images that produces the target; that is, $C\mathbf{x} = \mathbf{t}$. In all other cases, we will choose the vector $\mathbf{x}$ that minimizes the length of $\mathbf{r}$.

A Collection of Stars

Getting more personal, let's find a linear combination of the library images that best approximates the following picture of the author. Which celebrity do you think will be used most (least) in the construction of my photo?

The dimensions of this photo are 600×900 pixels. Each celebrity photo has the same dimensions. So, we are interested in solving:

$$C\mathbf{x} \approx \mathbf{t},$$

where C is a $540{,}000 \times 16$ matrix, $\mathbf{x}$ is a 16×1 vector and $\mathbf{t}$ is a $540{,}000 \times 1$ vector. That's a lot of numbers, so a computer will be helpful! Our goal is succinctly stated as choosing $\mathbf{x}$ that minimizes the length of the residual vector **r**. How do we find the vector **x** that produces this minimal length among the infinite possible values for the entries of **x**?

One approach is to solve the linear system

$$C^T C \mathbf{x} = C^T \mathbf{t}. \tag{10.3}$$

It may be surprising that solving this equation produces our desired **x** among the infinite number of possibilities. Why does it work? It's a bit beyond the scope of this book to unveil the theory behind why this is true. You may want to look it up. This method is called a *least squares problem*. The theory behind this approach to finding the solution, while requiring only a few lines of mathematics, involves taking the derivatives of the linear system $C\mathbf{x} = \mathbf{t}$ with respect to the elements of the vector **x**.

It turns out that solving equation 10.3 isn't actually the best method for solving such a problem on a computer. It can lead to considerable round-off error. A more reliable answer results from a method that involves the computation of a matrix decomposition. Such a method produced the results of this chapter.

So, for my image what combination of photos in the library produces the best result? See the results in Table 10.1.

TABLE 10.1.
Weights of the celebrity photos for the author's image.

Weight	Image		Weight	Image
0.342	Brad Pitt		0.013	Hugh Jackman
0.329	George Clooney		-0.019	Justin Bieber
0.273	Orlando Bloom		-0.020	Bruce Lee
0.222	Jennifer Lawrence		-0.054	Audrey Hepburn
0.134	Daniel Day Lewis		-0.075	Clark Gable
0.044	Tom Cruise		-0.089	Robert Downey Jr.
0.027	Neil Patrick Harris		-0.120	Ellen Degeneres
0.027	Leonardo DeCaprio		-0.155	Simon Cowell

Notice that the celebrities with the highest weights to produce an approximation to my image are Brad Pitt, George Clooney, and Orlando Bloom, all of whom have been named *People* Magazine's Sexiest Man Alive. I must say that I'm pretty pleased with the results coming from that residual vector of minimal length! Notice the next highest weight, Jennifer Lawrence, is given a weight just less than that of Orlando Bloom. Moreover, notice how several celebrities have negative weights, which corresponds to the contents of such pictures being subtracted in the sum of the images. So, the best approximation to my picture actually involves subtracting out 0.155 of Simon Cowell's image.

Why were such choices made? Keep in mind that we are approximating the solution to a linear system and not really facial features. The algorithm isn't trying to match noses or eyes. It simply matches, as best it can, numbers in a vector that result from matrix-vector multiplication. As such, the placement of the image plays an important role in the outcome. To see this, let's shift my image down slightly as seen in Figure 10.4. Now, the top 4 weights of the celebrity photos can be seen in Table 10.2. Note how Justin Bieber jumped from a negative weight for the image in Figure 10.4 (a) to the third highest weight. Further, my "connection" with George Clooney is lost at least in the top 4 celebrity weights.

(a)　　　　　　　　　　　　　(b)

Figure 10.4. The image in (a) is slightly shifted to create image (b).

TABLE 10.2.
Weights for the top 4 celebrity photos to create the best approximation of the author's
image in Figure 10.4 (b).

Weight	Image
0.351	Brad Pitt
0.279	Orlando Bloom
0.154	Justin Bieber
0.134	Jennifer Lawrence

 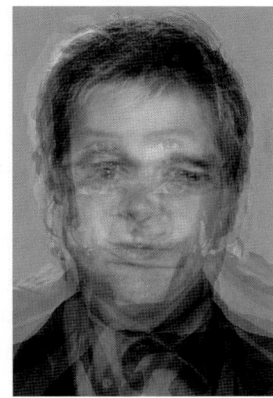

Figure 10.5. The original image (left) and the best approximation (right) using a
library of 16 celebrity photos.

To further see why the rankings are as they are, it helps to see the
image that results from our approximation. In Figure 10.5, we see the
best approximation (right) to the author's image (left). Notice how the
author's hair and dark shirt appear to be fairly well approximated. With
this in mind, look carefully at the images of Brad Pitt, George Clooney,
Orlando Bloom, and Jennifer Lawrence in our library of photos. It may
be easier to see why such stars topped the list.

Can we get a better image that is closer to the author's likeness?
Probably but it would require more photos. Further, the images in our
library differ in a variety of attributes such as the orientation of the face.
How can we find more uniform libraries of faces? It is relatively easy as

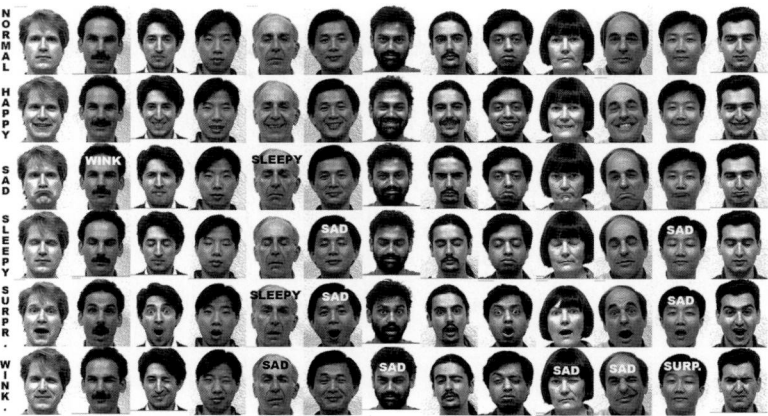

Figure 10.6. Faces from the Yale Face Database.

databases for facial recognition exist; some are free and others require a fee. You may want to use a search engine to find such resources. One of which is the Yale Face Database, which contains 5760 single light source images of 10 subjects each seen under 576 viewing conditions (9 poses x 64 illumination conditions). Such databases generally do not have celebrities as their subjects and would not enable you to describe your appearance in terms of a combination of famous faces. For instance, I could now describe myself as a combination of Brad Pitt, George Clooney, Orlando Bloom, and, yes, Jennifer Lawrence.

A more common method in facial recognition uses principle component analysis. Such an approach utilizes the eigenvectors of the covariance matrix resulting from the library of images and generally involves a library of several thousand images. In this context, the eigenvectors of the covariance matrix are often referred to as *eigenfaces*. We see such a set of eigenfaces for our library of celebrities in Figure 10.7.

Celebrity in Disguise

Let's change gears for a moment and disguise our celebrity. In particular, George Clooney decides to disguise himself with Hugh Jackman's goatee as seen in Figure 10.8 (left). Using this disguised image as

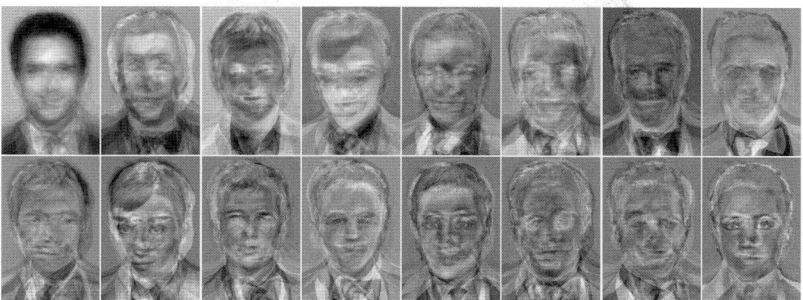

Figure 10.7. Eigenfaces for our library of famous faces.

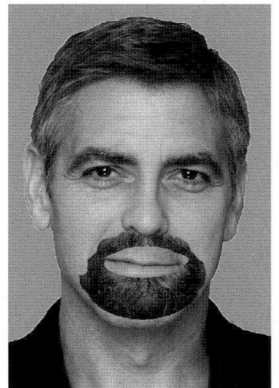 

Figure 10.8. A disguised George Clooney (left) and our algorithm's approximation (right).

our target image, what linear combination results from the library of celebrity images? Our algorithm chooses the following top 4 weights to produce the best approximation seen in Figure 10.8 (right).

Weight	Image
0.810	George Clooney
0.127	Hugh Jackman
0.086	Leonardo DeCaprio
0.079	Brad Pitt

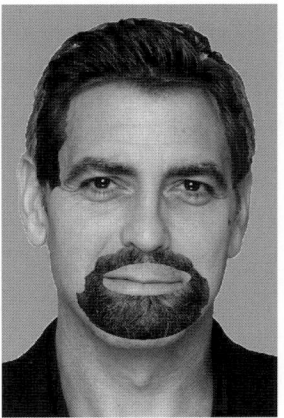 

Figure 10.9. A further disguised George Clooney (left) and our algorithm's approximation (right).

Wanting to be more mathematically incognito, suppose Clooney adds a toupee resembling Hugh Jackman's hairstyle. Can he fool our program now? In Figure 10.9, you see our algorithm's approximation (right) to the original image (left); what do you think?

While our algorithm's best approximation does indeed pick up the increased Jackman-ness of the photo, it still recognizes the dominant contributions of Clooney's photo. This is reflected numerically by the top 4 weights in the following table.

Weight	Image
0.743	George Clooney
0.156	Hugh Jackman
0.117	Tom Cruise
0.101	Leonardo DeCaprio

George Clooney will need to do better to fool our algorithm! Note, you could use this technique on your photo. You could, as with my photo, see what stars best approximate your image. If you want to change the results, alter your image with a new haircut, for instance, and see if you find a celebrity combination to your liking. In this sense, you can preview any potential mathematical makeover.

Problems from 200 BC

To end the chapter, here are two problems, from the *Jiuzhang Suanshu*, mentioned earlier in the chapter. Here's the first:

There are three classes of grain, of which three bundles of the first class, two of the second, and one of the third make 39 measures. Two of the first, three of the second and, one of the third make 34 measures. And one of the first, two of the second and three of the third make 26 measures. How many measures of grain are contained in one bundle of each class?

For this second problem write the associated linear system with five equations and six unknowns corresponding to this problem. The linear systems we solved in this chapter had no solution, which is why we used least squares. This puzzle has infinitely many solutions. Find the smallest solution in which the lengths are positive integers for the following problem from the ancient Chinese text.

There are five families which share a well. 2 of A's ropes are short of the well's depth by 1 of B's ropes. 3 of B's ropes are short of the depth by 1 of C's ropes. 4 of C's ropes are short by 1 of D's ropes. 5 of D's ropes are short by 1 of E's ropes. 6 of E's ropes are short by 1 of A's ropes. Find the depth of the well and the length of each rope.

> **Challenge 10.1.** *Find the solution to both of these problems from the Jiuzhang Suanshu.*

11. . . . .

March MATHness

THE MADNESS SWEEPS THE COUNTRY, infecting millions. Be careful. It happens every March and breaks out in many offices. We are talking March Madness, which is the Division I NCAA Men's basketball tournament. Whether you participate or not, many people do. In 2013, the ESPN online pool alone pitted over 8 million brackets against each other.

If you're immune, here's a quick tutorial. A March Madness bracket starts with match-ups of 64 teams in 32 games. For instance, Figure 11.1 is one quarter of a bracket in 2013. As you see, UNLV played California and Syracuse played Montana in the first round. Bracketology, as it is often called, involves deciding who would win those games. After you predict winners for the first 32 games, you decide who would win your predicted second round match-ups. You keep predicting winners in your bracket until you've projected a national champion.

March Madness pools award points to brackets for correct predictions. Pools vary in their scoring. ESPN's online pool awards 10 points for each correct choice in the first round, and each successive round doubles in the points allotted to a correct prediction. A bracket wins a pool by scoring more points than any other bracket.

Easy? Yes and no. Filling out a bracket can be as simple as flipping a coin to decide your projected winner. Producing a winning bracket

Figure 11.1. A portion of the 2013 March Madness bracket that, when complete, would predict your choice for the East Regional champion.

is often difficult; producing a bracket that correctly predicts every outcome in the tournament is nearly impossible. In fact, even with millions and millions of brackets submitted each year, there has never been a perfect bracket submitted to ESPN, CBS, or Yahoo! Sports—ever, not just last year.

Keep in mind, there are 9,233,372,036,854,775,808 (which is said 9 quintillion) ways to fill out a 64-team bracket. Let's take the 2012 population of the United States, 313 million. If every person in the U.S. produced 29.5 million different brackets and no two people had any brackets in common, only then would every possible bracket be covered. There's an issue. Even if you could fill out one bracket per second it would take over 900 years to create 29.5 million. The number of possible brackets reduces when you consider teams that should beat others. Therein lies the Madness. It's determining who beats who that no one has perfectly predicted!

Figure 11.2. In 2013, were the Louisville Cardinals destined to beat the Wichita State Shockers by the mere fact that their mascot is an animal?

Flip Out

With so many possibilities, the task of picking a bracket can be daunting. Given the many techniques people use and that none have produced a perfect bracket, why not be totally random? Maybe any team can win on any given day. With this in mind, let's construct brackets by determining each game's winner by the flip of a coin. We'll repeat the process several times. Five random brackets scored 320, 330, 220, 320, and 280 points in the ESPN Tournament Challenge in 2013. But note, the median score of 320 points would have beat the score of only 1.7% of the over 8 million brackets submitted to the online pool.

Picking Out a Pattern

The tournament isn't entirely random, even with the madness of unexpected results. For example, every team seeded #1 has won its first round game. So, giving this a 50-50 chance through the flip of a coin isn't wise. What other patterns might we find?

Here are some. Having a human mascot for the national championship team has been 10 times more likely than having a reptile. Championship teams have had blue in their uniform just over 55% of the time. Having red occurred just under 30%. Based on these types of patterns, were the Wichita State Shockers destined to lose to the Louisville Cardinals since 8 of the previous 10 teams in the championship game were teams with animal mascots? However, before 2013, every national champion since 2004 had had blue in its color scheme. How did Louisville (which has red and white in its color scheme) beat the University of Michigan (which has maize and blue)? Be careful assuming patterns in March Madness will always hold. If they did, we'd see perfect brackets.

From our coin flipping, we know total randomness doesn't help. We must analyze the right data in the right way.

Taking a Percentage

Let's try using data from the games played up to the March Madness tournament to create our brackets. We'll rank the teams using this data and then assume the higher ranked team wins. The question becomes how to rank the teams. Let's compute winning percentage, which equals the number of wins divided by the total number of games played. This produces a rating for the teams, and the higher the rating the better the team, at least mathematically. Who's number 1? The team with the highest rating. Note that Major League Baseball determines who plays in the playoffs by winning percentage. So, let's see how it can do for postseason basketball.

Let's rate all Division I NCAA Men's basketball teams by their winning percentage in games played against other Division I teams. The resulting bracket assumes the team with the higher winning percentage wins a game. This method predicts that Pittsburgh, an eighth seed would win the entire tournament; they lost in the first round to Wichita State, a team that made it to the Final Four. Moreover, the best teams in Division I men's basketball, as ranked by winning percentage, were Penn State, Texas Tech, and Hawaii, none of which made the tournament.

Still, these are isolated insights. Again, no bracket will be perfect so maybe this can still hit a home run in an office pool. To get a sense of this bracket's accuracy, let's submit it to the ESPN Tournament

Challenge, which contained over 8 million brackets in 2013. How does winning percentage method fare? The associated bracket received a score of 320 in the ESPN online pool. Note, two of the brackets created earlier by flipping a coin received this score. And, over 98% of the brackets submitted to ESPN's online pool would perform better.

Have we picked the wrong data to analyze for our bracket?

Following the Seed

A team that beats a lot of strong teams is different from a team that can only beat weak teams. Winning percentage doesn't differentiate between these types of wins. As such, a team from a much weaker conference can be rated high by winning percentage. March Madness considers this in the first round matchups. The teams are seeded with their performance in mind, which doesn't always align with their winning percentage.

So, let's create a bracket in which the higher seeded team wins. Returning to Figure 11.1, Indiana, NC State, UNLV, Syracuse, Butler, Marquette, Illinois and Miami would all win their first round game. In this tournament, this method didn't predict UNLV and NC State's losses. To be sure you're understanding the method, verify that this method would predict Indiana, Syracuse, Marquette, and Miami to win in the next round.

Note, this method is easy to implement until we get to the Final Four. Remember, Figure 11.1 is only a quarter of the tournament. There are 3 other regions, each containing a #1 seed. Our current method would predict that Louisville, Gonzaga, Kansas, and Indiana, all #1 seeds in the brackets for their regions, would reach the Final Four. At this point, let's predict the team with the higher winning percentage wins.

If we create an entire bracket using this method for the 2013, it would have scored 640 points in the ESPN online pool and beat 49.8% of the brackets. This is a significant jump from our earlier brackets that beat only 1.7%. Still, over half of the brackets are beating our method. Can we make another leap in accuracy?

Strength Building

Let's further build on the concept of a team's strength and rewarding wins against strong teams. Now, we will use linear equations to do this. In fact, the Bowl Championship Series, which ranks college football teams for the bowl games, has used such equations in its ranking methods specifically because they incorporate a sense of the strength of an opponent. One such method that utilizes linear equations is used by Ken Massey, who has served as a consultant for the BCS. His method started as an undergraduate honors math project and eventually made its way into the BCS [24]. The idea is to use least squares, which is the method that helped us determine which celebrity I look most like in the last chapter.

Here's the idea. Let's suppose some minor league baseball teams with unique names play each other. Suppose the Iola Gasbags beat the Pensacola Blue Wahoos by 2 runs. The Pensacola Blue Wahoos beat the Walla Walla Walla Wallas by 1 run. Do we then know that the Gasbags will be the Walla Wallas by 3 runs? If this were true, folks would go out to the ballpark for the peanuts and cracker jack and not the game. This type of transitivity will rarely, if ever, hold perfectly. Could we assume that it holds approximately? The Massey method does. Let r_1 be the rating for the Gasbags, r_2 the rating for the Blue Wahoos, and r_3 the rating for the Walla Wallas. The Massey method assumes these ratings can predict the outcomes of games. So, $r_1 - r_2 = 2$, since the Gasbags beat the Blue Wahoos by 2 runs, and $r_2 - r_3 = 1$. Let's add an additional game and assume The Walla Wallas beat the Gasbags by 4 runs, then $r_3 - r_1 = 4$.

This gives us 3 equations. However, there are no values for r_1, r_2 and r_3 that will make all 3 equations true. Let's construct the linear system, $M\mathbf{r} = \mathbf{p}$, or

$$
\begin{pmatrix} 1 & -1 & 0 \\ 0 & 1 & -1 \\ -1 & 0 & 1 \end{pmatrix} \begin{pmatrix} r_1 \\ r_2 \\ r_3 \end{pmatrix} = \begin{pmatrix} 2 \\ 1 \\ 4 \end{pmatrix}.
$$

Note, if teams play additional games, you simply add more rows to M and $\mathbf{p}$. Often, M will have many more rows than columns.

To solve this system, multiply both sides of the equation by the transpose of M, in which the rows become columns. So,

$$\begin{pmatrix} 1 & 0 & -1 \\ -1 & 1 & 0 \\ 0 & -1 & 1 \end{pmatrix} \begin{pmatrix} 1 & -1 & 0 \\ 0 & 1 & -1 \\ -1 & 0 & 1 \end{pmatrix} \begin{pmatrix} r_1 \\ r_2 \\ r_3 \end{pmatrix} = \begin{pmatrix} 1 & 0 & -1 \\ -1 & 1 & 0 \\ 0 & -1 & 1 \end{pmatrix} \begin{pmatrix} 2 \\ 1 \\ 4 \end{pmatrix},$$

which becomes

$$\begin{pmatrix} 2 & -1 & -1 \\ -1 & 2 & -1 \\ -1 & -1 & 2 \end{pmatrix} \begin{pmatrix} r_1 \\ r_2 \\ r_3 \end{pmatrix} = \begin{pmatrix} -2 \\ -1 \\ 3 \end{pmatrix}.$$

This system has infinitely many solutions. So, we take one more step and replace the last row of the matrix on the righthand size of the equation with ones and the last entry in the vector on the righthand side with a 0. This will enforce that the ratings sum to 0. Finding the ratings corresponds to solving the linear system

$$\begin{pmatrix} 2 & -1 & -1 \\ -1 & 2 & -1 \\ 1 & 1 & 1 \end{pmatrix} \begin{pmatrix} r_1 \\ r_2 \\ r_3 \end{pmatrix} = \begin{pmatrix} -2 \\ -1 \\ 0 \end{pmatrix},$$

which produces the desired ratings. This system gives the ratings $r_1 = -2/3, r_2 = -1/3$, and $r_3 = 1$. So, the ranking for this group of teams, from first to last, would be Walla Wallas, Blue Wahoos, and the Gasbags.

What difference does this make in creating a bracket for March Madness? Suppose we rate every Division I NCAA Men's basketball team and then create a bracket from these ratings. In 2013, the bracket created with the Massey method beat over 73% of the over 8 million brackets submitted online to ESPN.

Such ratings methods vary in their performance from year to year. So, you'll never know how such a ranking system will do at predicting the tournament. The math can advise you but isn't your crystal ball. You may want to combine the math with your own decisions to create a bracket and hope for the best! In the end, your bracket probably won't be perfect—but that's inevitably the madness that captures our attention.

12 . . . .

Ranking a Googol of Bits

TODAY, A NATURAL PLACE TO TURN to answer one's questions is the Internet. We might turn to Google, for instance, to search on the meaning of the word googol, which is the number 10^{100} after which Google was named, reflecting the company's goal of organizing all information on the World Wide Web (WWW). How do search engines rank web pages? What are some of the math and computing issues involved in returning web pages in an order deemed relevant to our query? You could search the internet for an answer, or keep reading.

Quality Control

Suppose that we submit the query computing to Google. An ordered list of web pages is returned. The top listed page is deemed the "best" page related to the query. Being at the top of the list is an enviable position. Why? Answer this: How many search engine results do you scan before selecting one? From the results of a poll that appeared in the *Wall Street Journal* replicated in Figure 12.1, we see that 39% of those polled scan only the first search result which was followed by 29% viewing only a few results. If you had a web page that you considered highly relevant to the word "computing" and it appeared on the fourth page of results

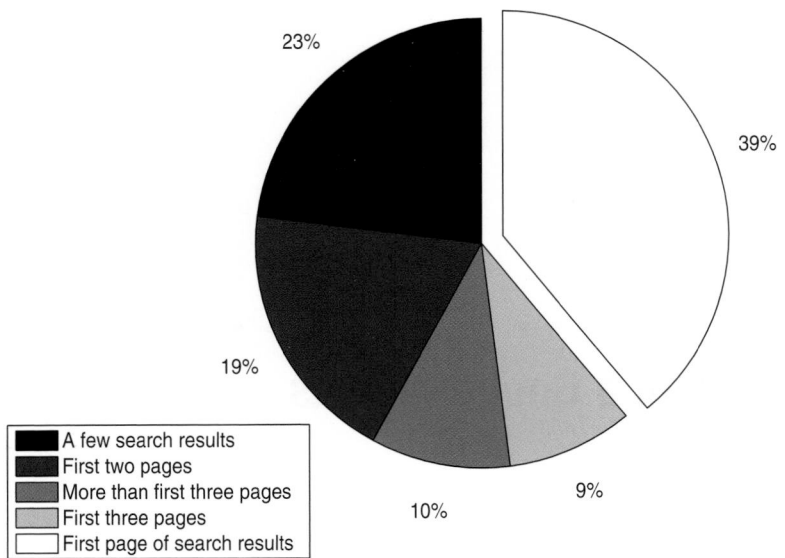

23%

39%

19%

A few search results
First two pages
More than first three pages
First three pages
First page of search results

10%   9%

Figure 12.1. Amount of Internet search results that web surfers typically scan before selecting one. (*Wall Street Journal*, April 13, 2007)

from a search engine, you should be concerned! Could you raise your ranking? Businesses do it by exploiting their knowledge of how search engines create such rankings.

Before discussing mathematical ideas imbedded in the algorithm that Google uses for rankings, let's motivate some of the ideas. What do we want to have returned from a query? First, we need web pages to be relevant to our query. We also need a sense of the quality of a web page. How do we do this? With billions of web pages out there, how do we possibly determine the quality of a page? Do we get everyone to vote? In a sense, yes. Feel slighted as you don't remember voting? If you have constructed a web page with any links, you actually did. If your web page has a link to a web page then, in a certain sense, your web page endorses the quality of that web page and supplies a vote regarding its quality.

Why? To answer this, let's generalize matters and consider interviewing job candidates. Suppose a job opening exists in our company and the following five candidates apply.

After some initial screening, we narrow our search to the final two candidates seen below.

For our final candidates, we look carefully at their references. The job application requested at least one reference and at most three. Our candidates supplied one and three references, respectively, as seen below:

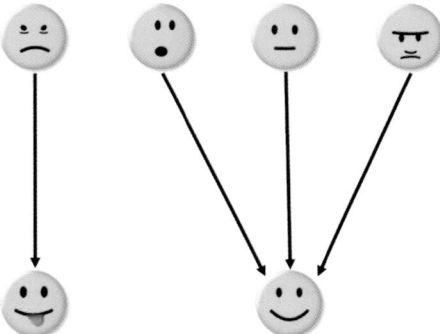

Who should we hire? Clearly, we should hire the candidate on the right. Why? There are three references. This person is clearly endorsed by more people and consequently better, right? Suppose, however, the candidate on the left has one very strong reference and the candidate on the right has 3 negative ones. Would this change your decision? The quality of the reference makes a difference in measuring the quality of the candidate.

Let's connect this to web pages. We could consider a web page as having high quality if it has a lot of links to it. However, it really depends on the quality of those links. Let's agree, at least for the moment, that a high quality website is www.CNN.com. Therefore, a link from www.CNN.com should raise the measure of quality of your

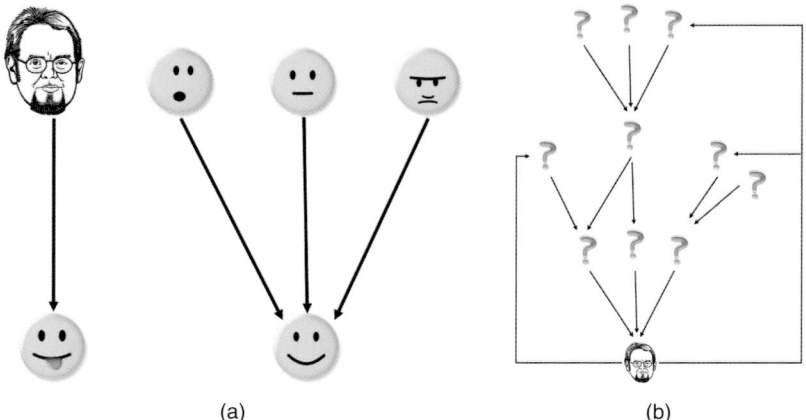

Figure 12.2. A network of references. In (a), Dr. Taylor is a reference for a job candidate. In (b), we consider a network of references related to Dr. Taylor.

web page. In a certain sense, it shouldn't be considered obscure if www.CNN.com considers it relevant. What if the link comes from a web page that is less well-known? Returning to our job candidates, what if that single reference to the job applicant comes from Dr. R. Taylor as pictured in Figure 12.2 (a), and you don't know Dr. Taylor or his field of expertise?

Since we don't know Dr. Taylor, we need to have a sense of the quality of his reference. One approach would be to look at his references. This necessitates knowing the quality of Taylor's references. So, we could ask the references for references. But, this would call for us to assess the quality of the references' references. Keep in mind that one such reference might just list Dr. Taylor as a reference. If this seems circular, that's because it is. Notice how we potentially need to know the quality of everyone as a reference at one time as pictured in Figure 12.2 (b).

Returning to web pages again, let's look at the small network in Figure 12.3. Each circle (or vertex) represents a web page, and an arrow (or edge) from web page i to web page j denotes a link on web page i to web page j. So, web page 1 has a link only to web page 4. Web page 4 has links to web pages 2, 3 and 5. Note, a link does not exist from web page 4 to web page 1 since the arrows have direction.

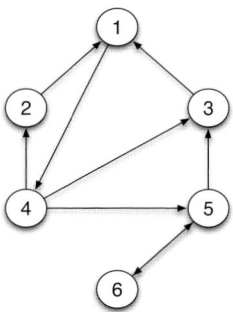

Figure 12.3. A small network of web pages.

We will consider web page 1 as endorsing web page 4. In a sense, web page 1 is recommending web page 4 for you to visit. While this is helpful, how much should we weight this endorsement to web page 4? It depends on the quality of web page 1. Web pages 2 and 3 endorse web page 1. Again, the weight of these recommendations depends on the quality of web pages 2 and 3. Notice how an analogous issue arises regarding the quality of recommendations for Figure 12.3 as for Figure 12.2 (b).

Google tackles this issue with the use of the PageRank algorithm, which was developed by Google's founders, Larry Page and Sergey Brin. The duo were graduate students at Stanford University when the foundational ideas of Google developed. The algorithm has two important attributes:

- the algorithm is determined entirely by the link structure of the WWW, and
- involves no content of any web page.

In the context of PageRank, it may be easier to think of a web page's popularity rather than its quality. This still connects with our idea of endorsement. If the most popular person in high school thought you were cool, your popularity could skyrocket, even if that was the only person, at least initially, who thought you were cool!

Google determines the popularity of a web page by modeling internet activity. If you visit web pages according to Google's model, which

Figure 12.4. Sergey Brin and Larry Page, founders of Google.

pages would you end up at the most? The frequency of time spent on a web page yields that page's PageRank.

What is Google's model of surfing? Is someone tracking your surfing to build the model? Google models everyone as being a random surfer by assuming that you randomly choose links to follow. In this way, the model stays away from making decisions based on preferred content. Let's look at this model by playing a game.

Playing Google-opoly in Monte Carlo

Do you have a fond memory of playing Monopoly and gaining sole ownership of the board's real estate market as your pile of funny money increases? Let's play a different game called Google-opoly introduced in [12]. The ideas behind this game led to billions of dollars, as they are some of the fundamental concepts behind PageRank. Page and Brin's version of Google-opoly is played on the entire World Wide Web. We'll play on much smaller game boards that involve only a few web pages.

We are about ready to play Google-opoly. The object of this game is to earn the most money by collecting from a surfer who pays one dollar to the owner of each web page he visits. Before explaining how to own web pages, let's take a closer look at a Google-opoly game board. Our game board, such as the one seen in Figure 12.3, will be a network of web pages represented as a directed graph. The board in Figure 12.3

TABLE 12.1.
Rules of Google-opoly for the network in Figure 12.3.

currently at page	roll a 1 visit page	roll a 2 visit page	roll a 3 visit page	roll a 4 visit page	roll a 5 visit page	roll a 6 visit page
1	4	4	4	4	4	4
2	1	1	1	1	1	1
3	1	1	1	1	1	1
4	2	2	3	3	5	5
5	3	3	6	6	6	6
6	6	6	6	6	6	6

contains six web pages. The directed edges indicate how we could move between the web pages by following a link on each page. For example, we see that web page 1 has a link to page 4, and web page 4 has links to pages 2, 3, and 5.

Understanding the board, we are ready for the rules of the game.

OBJECT: Own the web page that is the first to collect $1,000 from the surfer.

SET-UP: Each node in the graph represents a web page. Players (from 1 to 6) each choose a web page to own.

RULES:

1. The surfer begins at web page 1.
2. Each player in turn rolls the die. As determined by the number that appears on the die, move the surfer to the next web page according to the rules in Table 12.1.
3. Whenever the surfer lands on a web page, the page's owner receives $1.
4. The game ends when a web page collects $1,000. If someone owns such a web page, that person wins the game. Note that it is possible for no one to win a given round of the game.

Let's play the game for a little bit to get a sense of the game as it unfolds. First, let's select web page 1 to own. While the game starts at web page 1, we don't receive a dollar for this as it could be seen as an unfair advantage. A few rolls of the die direct the surfer over this network from web page 1 in the following sequence: 4, 2, 1, 4, 3, 1, 4, 5, 3. At this stage of the game, we have $2 but web page 4 has $3. Right now, web page 4 seems to have been a better choice to own, which is true for this set of rolls. Still, we have a long way to go for one of the web pages to collect $1,000.

Your turn. Choose a web page to own. Roll a die and see what unfolds. After 20 rolls how is your web page doing? Do you think you will win? What about after 40 rolls?

Depending on your sequence of rolls, you might not have even needed 20 turns to know if you would win! Once a surfer visits web page 6, no other web page will be visited. Such a web page with no outlinks is called a *dangling node*. So, owning web page 6 essentially guarantees that you will win a round of Google-opoly for this game board. With a proper outlook, you could make this winning conclusion before a single roll of the die. In fact, we can choose a winning web page for any board that contains one dangling node. This is obviously something that needs to be fixed before Google-opoly is a fair (and even fun) game. Before we figure out how to fix it, let's look at another game board to see if there are any more problems.

Let's play on the game board in Figure 12.5 that does not contain a dangling node. Playing with this new board necessitates adjustments to Table 12.1, which dictates where to go after each roll. Google-opoly assumes that a surfer is equally likely to visit any link on a web page. If you are at web page 1, then you must visit web page 4. Therefore, all rolls should lead from web page 1 to web page 4. Now, web page 4 has 3 links. So, there is a 1/3 chance of visiting any of its linked web pages when a hyperlink is followed from web page 4. In order to match this probability, rolling a 1 or a 2 will result in moving from web page 4 to web page 2. Can you think what entries would work for the remaining parts of the game? Keep in mind that in many cases the answers aren't unique. For instance, we could have set a rule that rolling a 3 or a 4 results in moving from web page 4 to web page 2. You will see one such set of adjusted rules in Table 12.2.

With the adjustments to the table complete, we are ready to play on the new board! So start rolling!

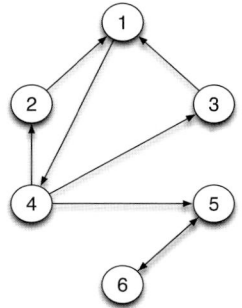

Figure 12.5. A second network for Google-opoly.

TABLE 12.2.
Rules of Google-opoly for the network in Figure 12.5.

currently at page	roll a 1 visit page	roll a 2 visit page	roll a 3 visit page	roll a 4 visit page	roll a 5 visit page	roll a 6 visit page
1	4	4	4	4	4	4
2	1	1	1	1	1	1
3	1	1	1	1	1	1
4	2	2	3	3	5	5
5	6	6	6	6	6	6
6	5	5	5	5	5	5

For this board, you should notice that you are guaranteed to eventually bounce back and forth between web pages 5 and 6 without any option of escape. This type of network structure is called a *cycle*. Keep in mind that a cycle could have more than 2 web pages. What is the winning guess? Web page 6 is a good guess but actually will acquire $999 at the point web page 5 collects $1,000 since web page 5 is always visited first and will be visited the same or one more time than web page 6.

So now we have played Google-opoly on two game boards. Is this what led to Brin and Page's billion dollar business? Actually, no. Part of the success of Google is that its rankings are based on a model of internet activity. Let's think more critically about our current rules for Google-opoly and how well they reflect randomly moving around the web.

Suppose you were surfing on the network in Figure 12.5 and enter a cycle. By the current rules of our game, any surfing session that lands on web page 5 gets stuck bouncing between pages 5 and 6. Alternatively, let's return to the issue of the dangling node in Figure 12.3. Our current rules have us stuck at that web page <u>forever</u>! Better be careful visiting a music or video file under this model as that would be your last page to visit. This behavior doesn't reflect how people actually surf the internet. When you reach a dangling node or enter a cycle, you either enter another web page's address or click the back button. So, we will introduce a new idea into our game that will lead to an altered set of rules that better reflect surfing behavior and also make for a more interesting game of Google-opoly.

Keep in mind that a surfer has two choices at any given page—follow a link or go to a different web page. Our rules of Google-opoly will now reflect these options. When we jump to a web page in our network, we'll assume that *every* web page in the network is an equally likely destination. Brin and Page called this jumping *teleportation*.

Let's play again on the network of web pages from Figure 12.3. We will now use two dice with different coloring as we will need to distinguish between them. We will refer to the die as Die 1 and Die 2. Note you could also simply roll one die twice. The rules of the game are now:

OBJECT: Own the web page that is the first to collect $1,000 from the surfer.

SET-UP: Each node in the graph represents a web page. Players (from 1 to 6) each choose a web page to own.

RULES:

1. The surfer begins at web page 1.
2. Each player will move the surfer in turn as determined by the roll of the dice as detailed in the following steps.
3. If the surfer is at a dangling node, then teleport by following these steps:

- Roll both dice.
- The number that appears on Die 1 is the next web page for the surfer to visit, which may be the same web page. Go to step 5.

4. To get to this step, the surfer is not on a dangling node. Roll both dice. This will determine whether the surfer follows a link on the current web page or teleports, and also the next web page to visit.

  - If you roll 1 through 5 with Die 2, then

    - the surfer follows a link on the current web page;
    - the value rolled by Die 1 indicates where to visit next according to the corresponding rule in Table 1.

  - If you roll a 6 with Die 2, then the surfer will teleport. In this case, the number that appears on Die 1 is the next web page to visit.

5. Whenever the surfer lands on a web page, the page's owner receives $1.
6. If no web page has collected $1,000 go to step 2.
7. The game ends when a web page collects $1,000. If someone owns such a web page, that person wins the game.

Your turn; start rolling.

Your game is produced with a random walk which will probably be more like a short meander over the board. To get a good approximation, we need to go on a long random hike through the web pages to get a better idea of the desired probabilities. In the bar chart in Figure 12.6, you'll see that web pages 1 and 4 were in a bit of a neck and neck race for the most revenue. With only $1,000 being the revenue that wins the game, either web page 1 or 4 might win. Even if web page 4 wins a particular round, if you allow the number of rolls to increase, the bar chart should settle down and indicate that web page 1 would receive the most revenue over more time. Depending on your sequence of rolls, this can take some time to observe. You can see one set of rolls in Table 12.3.

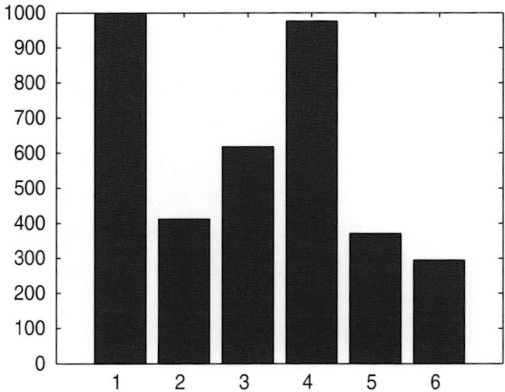

Figure 12.6. Bar chart for one game of Google-opoly.

TABLE 12.3.
Probability of visiting each web page in Figure 12.3 by rolling a die.

Number of Simulations	Page 1	Page 2	Page 3	Page 4	Page 5	Page 6
2000	544	215	318	518	221	185
4000	1079	436	640	1042	437	367
6000	1618	664	954	1561	664	540
8000	2167	905	1281	2088	867	693
10000	2706	1132	1589	2618	1091	865

Rolling with Markov

Google-opoly involves computing the total number of times each web page is visited at the current stage of the game. Suppose instead, the histograms reported the proportion of time each web page had been visited. Calculating such proportions if Google-opoly were played forever computes the *PageRank* for the web pages in that network. The ability of PageRank to give useful information is a key part of the success of Google's search engine.

Censoring Google

In 2005, Google engineered a search engine specifically for China (www.google.cn). Most countries use a version of Google that is identical to www.google.com except that the text is in the native language of that country rather than in English. The Chinese version of Google met the censorship requirements of the Chinese government. However, in 2010, Google announced that they were no longer willing to censor searches in China, which led to the business ending some of its services to China.

How does Google find the PageRank of the billions of web pages that it indexes? Clearly, the company doesn't pay its employees to roll multi-billion-sided dice. Does a computer program simulate the game of Google-opoly and randomly surf over the network? The answer is no, but the technique that is used produces equivalent results. Instead of simulating a random surfing session, we will create a matrix G in which the entries g_{ij} equal the probability of being at web page i and moving to web page j.

Let's compute the first row of G for the network in Figure 12.3. Entries in this row correlate to a probability of jumping from web page 1. The entry in the first column, g_{11}, equals the probability of moving from web page 1 to web page 1. This only happens through teleportation. In our game, 1/6 of the time we teleport, and at that time every page is an equally likely destination. So, each page has a (1/6) (1/6) = 1/36 chance of being visited through teleportation from web page 1. In the game, we follow hyperlinks on a page rather than teleport 5/6 of the time. Web page 1's only link is to web page 4. So, web page 4 will be visited through teleportation 1/36 of the time and through web page 1's hyperlink 5/6 of the time. So, there is a 31/36 chance of visiting web page 4 after your first turn of Google-opoly. Therefore, the first row of G is

$$\mathbf{g}_1 = (1/36 \ \ 1/36 \ \ 1/36 \ \ 31/36 \ \ 1/36 \ \ 1/36).$$

Before producing the entire matrix, let's produce the sixth row which assumes you are at the dangling node. So, we must teleport. Since all web pages are equally likely as a destination, each web page has a 1/6 probability of being visited from web page 6. So, the sixth row of G is

$$\mathbf{g}_6 = (6/36 \ \ 6/36 \ \ 6/36 \ \ 6/36 \ \ 6/36 \ \ 6/36).$$

Similar logic for all the rows produces

$$G = \begin{pmatrix} 1/36 & 1/36 & 1/36 & 31/36 & 1/36 & 1/36 \\ 31/36 & 1/36 & 1/36 & 1/36 & 1/36 & 1/36 \\ 31/36 & 1/36 & 1/36 & 1/36 & 1/36 & 1/36 \\ 1/36 & 11/36 & 11/36 & 1/36 & 11/36 & 1/36 \\ 1/36 & 1/36 & 16/36 & 1/36 & 1/36 & 16/36 \\ 6/36 & 6/36 & 6/36 & 6/36 & 6/36 & 6/36 \end{pmatrix},$$

which is called a *Markov transition matrix*. Our game states that we always start at web page 1. The vector

$$\mathbf{v} = (1 \ \ 0 \ \ 0 \ \ 0 \ \ 0 \ \ 0),$$

like the matrix G contains probabilities as its elements. Since the first entry is 1, the vector states that you are, with probability 1, at web page 1. To determine the probabilities of visiting the web pages after one turn, compute

$$\mathbf{v}G = \mathbf{v}_1 = (1/36 \ \ 1/36 \ \ 1/36 \ \ 31/36 \ \ 1/36 \ \ 1/36).$$

The probabilities of visiting the web pages after two turns are

$$\mathbf{v}_1 G = \mathbf{v}_2 = (0.07793 \ \ 0.27083 \ \ 0.28241 \ \ 0.05478 \ \ 0.27083 \ \ 0.04321).$$

Note that

$$\mathbf{v}_2 = \mathbf{v}_1 G = \mathbf{v}G^2. \text{ So, } \mathbf{v}_{100} = \mathbf{v}G^{100}.$$

If we compute $\mathbf{v}_n$ for increasingly large n, we converge to the steady-state vector. For instance, for this example, $\mathbf{v}_{50} = (0.26648\ 0.11260\ 0.15951\ 0.26198\ 0.11260\ 0.08674)$ is the same to five decimal places as $\mathbf{v}_{60}$. From the entries, we see that 26.6% of the time, if we surf forever, we will visit web page 1. We have found the best web page to own in Google-opoly for this network. You can compute similar steady-state vectors for the other game boards. Notice how we found the PageRank of the network with $\mathbf{v}_{50}$ using the Markov transition matrix, whereas we would need more than 50 (a lot more) rolls of the dice to find the same PageRank values for the system were we to randomly surf.

At the core of Google's rankings is the PageRank algorithm created by Page and Brin. The Google-opoly game is essentially this method. Keep in mind that winning Google-opoly only requires determining the top ranked web page. PageRank, as it should, ranks all the web pages in the network. Page and Brin's billion dollar business of Google included more ideas than that found in Google-opoly. For example, when you submit a query, the list of web pages that Google returns results from a page's ranking but also a measure of how relevant the page's content is to the topic of your query. Even so, understanding Google-opoly introduces an important part of Google and its success.

Theoretically Speaking

Let's look at PageRank at a bit of a deeper level. Again, we were finding the steady-state vector where $\mathbf{v} \approx \mathbf{v}G$. PageRank is based on setting this to equality which means $\mathbf{v}G = \mathbf{v}$. Google defines the PageRank of page i to be the ith element of such a vector $\mathbf{v}$, which is sometimes called the *PageRank vector*. Ah, but $\mathbf{v}G = \mathbf{v}$ is only true if we truncate the values of $\mathbf{v}$ to 5 decimal places when we look at $\mathbf{v}_{50}$ and $\mathbf{v}_{60}$. Truncating to 5 decimal places can distinguish the rankings for this small system. Given that Google indexes billions of web pages, we may need convergence to more than four decimal places. Maybe we simply didn't print enough significant digits.

What if we want say 16 decimal places? Do we keep finding G^k for larger and larger k? This is one, very efficient approach. Another outlook leans on the fact that a vector $\mathbf{v}$ that satisfies the property $\mathbf{v}G = \mathbf{v}$ is called the left eigenvector of G.

Usually linear algebra classes deal with right eigenvectors. For instance, if we had a matrix A and a vector $\mathbf{x}$ such that $A\mathbf{x} = -2\mathbf{x}$, then $\mathbf{x}$ is a right eigenvector of A corresponding to the eigenvalue -2. If $\mathbf{y}A = 5\mathbf{y}$, then $\mathbf{y}$ is a left eigenvector of A corresponding to the eigenvalue 5. In this example, the number 5 is sometimes called a left eigenvalue, but a nice fact about matrices is that the right and left eigenvalues of a matrix are the same. So, we can simply call 5 an eigenvalue of A.

Let's look at a quick example. Suppose

$$A = \begin{pmatrix} 5 & -7 & 7 \\ 4 & -3 & 4 \\ 4 & -1 & 2 \end{pmatrix}$$

Then,

$$A \begin{pmatrix} 1 \\ 1 \\ 1 \end{pmatrix} = \begin{pmatrix} 5 \\ 5 \\ 5 \end{pmatrix} = 5 \begin{pmatrix} 1 \\ 1 \\ 1 \end{pmatrix}.$$

So, we know that 5 is an eigenvalue of A. Be careful not to assume too much information from our knowledge of the associated right eigenvector. Notice that

$$(1\ 1\ 1)A = (13\ -11\ 13).$$

If we want to find the left eigenvector associated with the eigenvalue 5, then we can find the eigenvectors of A^T. This would lead us to see that:

$$(-1\ 1\ -1)A = (-5\ 5\ -5).$$

So, in this example, the eigenvalue 5 has different left and right eigenvectors.

Why all this about eigenvectors? Looking at the PageRank vector as an eigenvector enables us to lean on the theory developed in the field of linear algebra to aid us in our computation. Is the PageRank vector unique? If not, which vector would you choose? Would you allow companies to bid on which vector you would use? It turns out that the PageRank vector is *always* unique regardless of the network of web pages. That's a strong statement. Mathematical theorems allow such absolutes to be known and stated. Here, we lean on a theorem known as Perron's Theorem.

> **Perron's Theorem** Every real square matrix with entries that are all positive has a unique eigenvector with all positive entries; this eigenvector's corresponding eigenvalue has only one associated eigenvector, and this eigenvalue is the largest of the eigenvalues.

Why can we use this theorem? First, the Google matrix G is a square matrix and all of its entries are real numbers. Note the last requirement for this theorem to be used–the matrix must have all positive entries. Teleportation guarantees that this will always be true for any G corresponding to any network!

So, Perron's Theorem applies to any Google matrix G. What does it tell us? There is a unique eigenvector with all positive entries. That is, one and only one left (or right) eigenvector of G has all positive entries and the eigenvalue associated with this eigenvector is the largest eigenvalue; this eigenvalue also only has one associated eigenvector. Let's see how this helps us.

It's easy to find an eigenvector of G with all positive entries. Recall that the rows of G sum to 1. Therefore, $G\mathbf{1} = \mathbf{1}$, where $\mathbf{1}$ is the column vector of all ones. That is, $\mathbf{1}$ is a right eigenvector of G associated with the eigenvalue 1.

Perron's Theorem ensures that $\mathbf{1}$ is the unique right eigenvector with all positive entries, and hence its eigenvalue must be the largest eigenvalue. This means that there must also be a unique vector $\mathbf{v}$ that satisfies $\mathbf{v}G = \mathbf{v}$. This vector must have all positive entries and will be

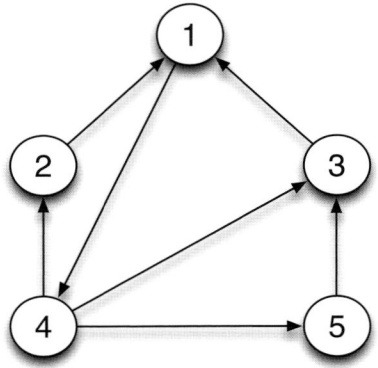

Figure 12.7. A third network of web pages.

unique. For a Google matrix G, the PageRank vector **v** is unique, and its entries can be viewed as probabilities for our desired ranking.

What did the math theory give us? It guarantees that no matter how much the web changes or what set of web pages Google indexes, the PageRank vector can be found and will be unique.

Let's close by finding the PageRank vector for the network in Figure 12.7. Solving $\mathbf{v}G = \mathbf{v}$ yields

$$\mathbf{v} = (0.2959 \ 0.1098 \ 0.2031 \ 0.2815 \ 0.1098).$$

So, a random surfer will visit page 1 approximately 27% of the time and page 2 about 11% of the time.

Note that Pages 1 and 3 have the same indegree and outdegree, yet page 1 is linked from pages that, in the end, have higher PageRank. Page 4 receives a high PageRank because it is linked from page 1. If a surfer lands on page 1 (which occurs about 29% of the time), then 85% of the time the surfer will follow a link. Page 1 links only to page 4. The high PageRank of page 1 boosts the probability that page 4 is visited!

Note that it pays to have a recommendation from a web page with high PageRank! There are companies that you can pay to boost your PageRank. This is accomplished in a variety of ways. You could try it yourself! Just figure out how to get a web page with high PageRank to link to your page. Get in the news, have CNN.com link to your page and who knows where you will land in the web results of Google.

Then again, be careful what you do to get in the news or you may be viewing those web results while spending your time in a small cell with no windows!

GOOGLE BOMBS

Much can be gained for those who can use knowledge of the underlying mathematics and computer science in information retrieval to their advantage. For example, in October of 2003, George Johnston initiated a *Google bomb* called the "miserable failure" bomb, which targeted President George W. Bush and detonated by late November of that same year. The result was Google returning the official White House Biography of the President as the highest ranked result to the query miserable failure. Putting together a Google bomb was relatively easy, it involved several web pages linking to a common web page with an agreed upon anchor text (the text that one clicks to go to the hyperlink). In the case of the "miserable failure" project, of the over 800 links that pointed to the Bush biography, only 32 used the phrase "miserable failure" in the anchor text [22]. This ignited a sort of political sparring match among the web saavy, and by January 2004, a "miserable failure" query returned results for Michael Moore, President Bush, Jimmy Carter, and Hillary Clinton in the top four positions. As expected, Google is fully aware of such tactics and works to outsmart these and other initiatives that can dilute the effectiveness of its results. For more information, search the internet on "Google bombs" or "link farms."

13...........................

A Byte to Go

WE'VE BITTEN OFF A LOT OF MATH. Are you creating a candy mosaic to top a cake? Maybe you're collecting photos to illuminate who you look like. Or, possibly you're admiring the results of the mathematics involved in the results to your queries in search engines. Whatever the case may be, we have digested a lot of math through the preceding pages of this book. As your mind is mathematically nourished, it may get energized with new ideas! Applying and adapting concepts is an important part of mathematics and computing.

It is now your turn. What ideas come to mind? What concepts in the book caught your attention? Can you think of other applications? Maybe you should do more reading to learn more of the mathematics behind a method. Maybe you can adapt an algorithm to improve its accuracy or efficiency. You may wish to see if there are more advanced methods to accomplish a task. Warm up your computer as the mathematical problems in this book often require computer bytes to solve, even on moderately sized problems. The bit of mathematics we've explored can serve as a springboard into a large array of exciting areas of mathematical study.

14......................

Up to the Challenge

CHALLENGE QUESTIONS HAVE APPEARED throughout the book. Below are answers to these questions although it is entirely possible that you devised another approach.

2.1. Note that if 3 divides the sum of the digits that comprise a number, then 3 divides that number. Now, $3+9+8+7 = 27$, which is divisible by 3. So, 3987 is divisible by 3. Since $4+3+6+5 = 18$, which is divisible by 3,4365 is also divisible by 3. This relationship works the other way, too. That is, a number is divisible by 3 if 3 divides the sum of the digits that comprise that number. So, 4472 is not divisible by 3 since $4+4+7+2 = 17$. Looking at the equation

$$3987^{12} + 4365^{12} = 4472^{12},$$

the left-hand side is divisible by 3^{12} but the right-hand side is not, so the equation cannot be true.

2.2. If Jay had purchased four of the BMWs for every two Rolls Royces, his idea to sum the two prices to get "six for $840,000" would have worked. However, because Jay purchased equal amounts of BMWs and Rolls Royces, he needed to add "four for $340,000" with "four for $1,000,000" to get "eight for $1,340,000" which would have gotten him $2,680,000 in sales as opposed to $2,240,000.

2.3. The first line states that $a = b$. Things progress fine until we cancel common terms, because we are dividing both sides by $(a - b)$, which equals 0. Dividing by zero has no meaning. Consider $c/0$, for c not equal 0. There is no number which, multiplied by 0, gives c and so division by zero is undefined. We see how in this problem it leads to problems!

3.1. Dimes take 19 days like pennies so the total equals $5,242.87 \times 10 = \$52,428.70$. Quarters take 17 days since $2^{16}(5.6/1000) = 367$. The total is \$32,767.75.

3.2. We start with 40 people and then 80, 160, and so forth. After 14 levels, the negative reviews would be tweeted by $40 * (2^{15} - 1) = 1,310,680$ people.

3.3. Since there are, on average, 81.9 characters per tweet and 5.5 characters on average in an English word, then 9,420 tweets per second would result in 771,498 characters or 140,272 words per second. This is more words than *Twilight* but less than *The Order of the Phoenix*. Note, if this rate continued for 10 seconds, there would be over 1 million words tweeted on the topic which would be more words than in all the books mentioned in Chapter 3.

3.4. Let's take the population of the Earth to be 7 billion. Now, if you start with $2^1 = 2$ and use this to compute $2^2 = 4$ and continue you'll find $2^{32} = 4,294,967,296$ and $2^{33} = 8,589,934,592$. So, we need at most 33 guesses to identify a name from all the names of people on the entire Earth, if they were ordered in a book of names.

5.1. Plotting the lines, you find the secret word to be:

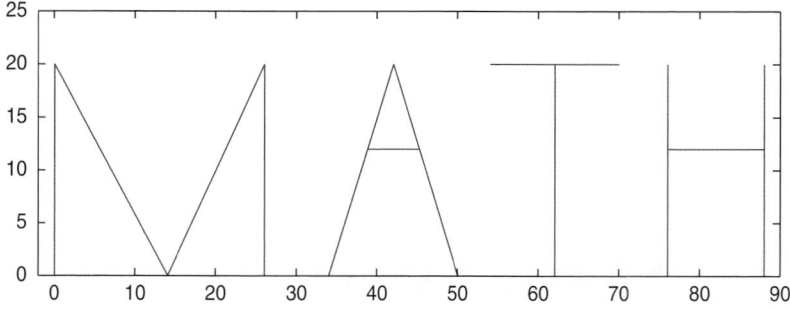

5.2. Careful analysis leads to the word:

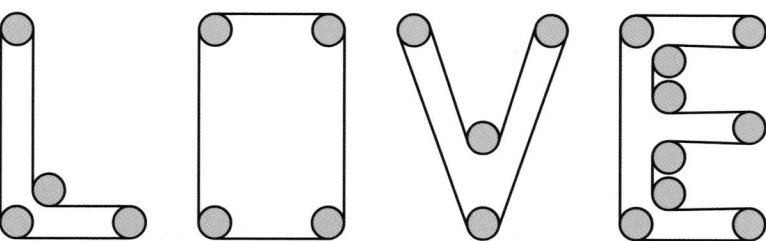

5.3. Recall from Chapter 5, a red Angry Bird's trajectory can be described by

$$y = 1 + \tan(\theta)x - \frac{1}{(4.35)^2 \cos^2 \theta} x^2.$$

So, a bird launched at 30 degrees will follow the path $y = 1 + 0.57735x - 0.0705x^2$. The vertex of the parabola occurs at $x = 0.57735/(2(0.0705)) = 4.09468$ at a height equaling $1 + 0.57735(4.09468) - 0.0705(4.09468)^2 = 2.182$, which is just over twice the height of the slingshot from the ground.

6.1. The Touching Numbers Puzzle can be challenging. Trial and error will lead to higher and lower scores for alternate arrangements. What is the largest score? Even if you found it, it might be difficult to know you indeed calculated the minimum value. The largest possible score for an arrangement is 352; the smallest is 181. What configurations lead to these values? Let's maintain the challenge of the puzzle until you solve it. At least now, you will know when you find the solution.

6.2. The key is finding the region that is colored red below. Then, you simply place the start and end of your maze on the boundaries of the red region.

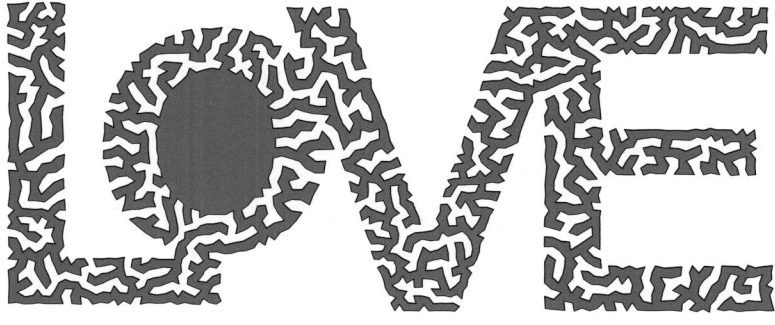

8.1. The 11 by 11 grid on the left leads to the chocolate chip mosaic on the right containing 83 milk chocolate chips and 21 white

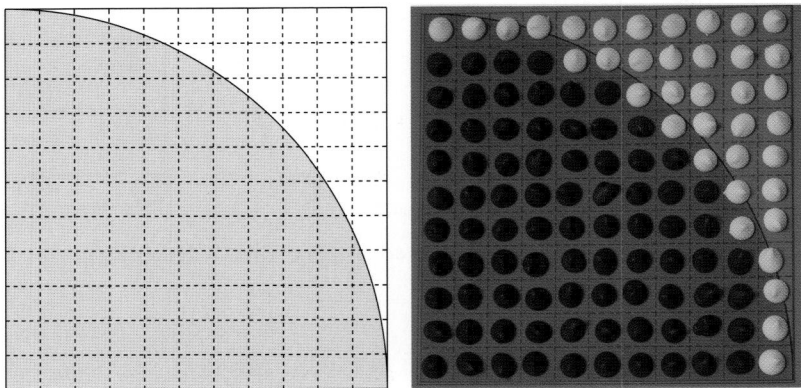

chocolate chips on squares with only a portion of the underlying circle. Therefore, $\pi \approx 3.09$, which is significantly improved from the estimate of 2.7438 found using the method in the chapter.

9.1. If we change the coloring of a vector $\mathbf{v}$ by performing $\mathbf{w} = B\mathbf{v}$. We can undo our recoloring with a matrix U by computing $U\mathbf{w}$ if $U = B^{-1}$. The matrix U is said to be an inverse of B. However, not all matrices have inverses and so not all colorings can be undone. For example, if we take

$$B = \begin{pmatrix} 1/3 & 1/3 & 1/3 \\ 1/3 & 1/3 & 1/3 \\ 1/3 & 1/3 & 1/3 \end{pmatrix},$$

then the resulting image is a grayscale version of the original. This cannot be undone. We can't find a matrix U such that if $\mathbf{w} = B\mathbf{v}$ then $\mathbf{v} = U\mathbf{w}$. Many colorings of an image would result in the same grayscale image.

10.1. The first problem taken from the *Jiuzhang Suanshu* leads to the linear equations

$$\begin{aligned} 3x + 2y + z &= 39 \\ 2x + 3y + z &= 34 \\ x + 2y + 3z &= 26, \end{aligned}$$

which has the solution 9.25, 4.25, and 2.75 for the measures of grain in the three classes.

The second problem asks for the depth of a well and the length of ropes, which leads to the linear equations:

$$2A + B = W$$
$$3B + C = W$$
$$4C + D = W$$
$$5D + E = W$$
$$6E + A = W$$

This system has 6 unknowns and 5 equations and has infinitely many solutions of the form

$$A = \tfrac{265}{721} W$$
$$B = \tfrac{191}{721} W$$
$$C = \tfrac{148}{721} W$$
$$D = \tfrac{129}{721} W$$
$$E = \tfrac{76}{721} W.$$

We will choose solution that leads to the shortest ropes where all the lengths are positive integers. This leads to the solution that ropes A, B, C, D, and E have lengths 265, 191, 148, 129, and 76. The well has a depth of 721.

BIBLIOGRAPHY • • • • • • • • • • • • • • • • •

[1] R. Allain, *The physics of Angry Birds*. http://www.wired.com/wiredscience/2010/10/physics-of-angry-birds/.

[2] M. Anderson, R. Wilson, and V. Katz, eds., *A History of Mathematics: An Introduction*, Addison Wesley, 2nd ed., 1998.

[3] M. W. Berry and M. Browne, *Understanding Search Engines: Mathematical Modeling and Text Retrieval (Software, Environments, Tools), Second Edition*, Society for Industrial and Applied Mathematics, Philadelphia, PA, USA, 2005.

[4] R. Bosch, *Constructing domino portraits*, in Tribute to a Mathemagician, B. Cipra, E. Demaine, M. Demaine, and T. Rodgers, eds., A.K. Peters, 2004, pp. 251–256.

[5] ———, *Opt art*, Math Horizons, (2006), pp. 6–9.

[6] R. Bosch and A. Herman, *Continuous line drawings via the traveling salesman problem*, Operations Research Letters, 32 (2004), pp. 302–303.

[7] R. Bosch and C. S. Kaplan, *TSP art*, in Proceedings of Bridges 2005, 2005, pp. 303–310.

[8] E. B. Burger and M. Starbird, *The Heart of Mathematics: An Invitation to Effective Thinking*, Wiley, 2nd ed., 2008.

[9] T. Chartier, *Googling Markov*, The UMAP Journal, 27 (2006), pp. 17–30.

[10] T. Chartier, D. Clayton, M. Navas, and M. Nobles, *Mathematical penmanship*, Math Horizons, (2008), pp. 10–11, 31.

[11] T. Chartier and D. Goldman, *Mathematical movie magic*, Math Horizons, (2004), pp. 16–20.

[12] T. Chartier, E. Kreutzer, A. Langville, and K. Pedings, *Goolge-opoly*, Loci, 1 (2009).

[13] M. Chlond, *The traveling space telescope problem*, INFORMS Transactions on Education, 3 (2002). http://ite.informs.org/Vol3No1/Chlond/.

[14] W. J. Cook, *In Pursuit of the Traveling Salesman: Mathematics at the Limits of Computation*, Princeton University Press, 2012.

[15] E. Demaine, M. Demaine, and B. Palop, *Conveyer belt font*. http://erikdemaine.org/fonts/conveyer/, accessed March 2013.

[16] M. Gardner, *Mathematics, Magic and Mystery*, Dover, 1956.

[17] ———, *Aha!: Aha! Insight and Aha! Gotcha*, Mathematical Association of America, 2006.

[18] A. Greenbaum and T. Chartier, *Numerical Methods: Design, Analysis, and Computer Implementation of Algorithms*, Princeton University Press, 2012.

[19] S. Greenwald and A. Nestler, *Mathematics and mathematicians on The Simpsons: Simpsonsmath.com*. http://www.simpsonsmath.com, accessed March 2013.

[20]   C. L. Grossman, *It's Tebow time: Denver quarterback inspires nation*, USA Today, (2012).

[21]   V. Katz, *A History of Mathematics: An Introduction*, Addison Wesley, 2nd ed., 1998.

[22]   A. N. Langville and C. D. Meyer, *Google's PageRank and beyond: the science of search engine rankings*, Princeton University Press, Princeton, NJ, 2006.

[23]   G. Leonhard, *Mediafuturist: My tweet-stats, and some cool Twitter stats & visualization tools.* http://www.mediafuturist.com/2009/06/some-very-cool-twitter-related-stats-via-tweetstats—this-shows-the-most-often-used-topics-and-keywords-that-i-used-in-m.html, accessed March 2013.

[24]   K. Massey, *Statistical models applied to the rating of sports teams*, undergraduate honors thesis, Bluefield College, 1997.

[25]   J. Nichols, J. Mahmud, and C. Drews, *Summarizing sporting events using Twitter.*

[26]   Y. Perelman, *Mathematics Can Be Fun*, Mir, 1985.

[27]   J. R. Pierce, *An Introduction to Information Theory: Symbols, Signals and Noise*, Dover, 1980.

INDEX •

IMAGE CREDITS ● ● ● ● ● ● ● ● ● ● ● ● ● ● ● ● ●

p. 2 (fig. 1.1): The Library of Congress

p. 3: The National Center for Computational Sciences, Oak Ridge National Laboratory

p. 9: BMW courtesy http://commons.wikimedia.org/wiki/File:2008-2010_BMW_120i_%28E88%29_convertible_%282011-11-08%29_01.jpg; Rolls Royce courtesy http://15pictures.com/15-pictures-rolls-royce-phantom/

p. 13 (fig. 3.1): https://en.wikipedia.org/wiki/File:VespasianDupondius.jpg

p. 14 (fig. 3.2): http://www.artpaintingsgallery.com/paintings/archives/1672/antonio-del-pollaiolo-hercules-and-the-hydra-and-hercules-and-anteo-painting

p. 16 (fig. 3.3): © Twitter

p. 18: Beyonce © s_buckley/ShutterStock.com; Tebow © s_buckley/ShutterStock.com

p. 22 (fig. 4.1): image by Nol Aders, © Creative Commons Attribution-Share Alike 3.0

p. 31: fig. 4.12a © Ken Musgrave; fig. 4.12b http://www.pandromeda.com/gallery/img.php?type=stilloriginal&id=cVplqp004&f

p. 33 (fig. 5.1): Scriptorium monk courtesy http://commons.wikimedia.org/wiki/File:Scriptorium-monk-at-work.jpg; illuminated letter © Zvonimir Atletic / Shutterstock.com

p. 36 (figs. 5.4–5.5): © Erik Demaine

pp. 37–40 (figs. 5.6–5.9): © Rovio

p. 41: Courtesy of Lucasfilm Ltd., LLC. *Star Wars: Episode II—Attack of the Clones* © & ™2002 Lucasfilm Ltd., LLC. All rights reserved. Used under authorization. Unauthorized duplication is a violation of applicable law.

p. 44: http://140.116.36.16/whoiswho/whoiswho.htm

p. 47 (fig. 6.7): Copyright © 2012 Hordern-Dalgety Collection. http://puzzlemuseum.com

p. 48: fig. 6.8a © Proctor and Gamble; fig. 6.8b courtesy William Cook

p. 55 (fig. 7.1): courtesy of Shepard Fairey / Obeygiant.com

p. 59 (fig. 7.7a): image by Inductiveload, © GNU Free Documentation Licence

p. 71 (fig. 8.11): courtesy of Shepard Fairey / Obeygiant.com

p. 71 (fig. 8.12): http://en.wikipedia.org/wiki/File:Abraham_Lincoln_head_on_shoulders_photo_portrait.jpg

p. 72: © Bob Bosch

pp. 73, 74: © W.A. Elliot Co. Toronto

pp. 77, 80, 81 (figs. 9.1, 9.4, 9.6): © Corbis

p. 82 (fig. 9.7): Matlab

pp. 83–84 (figs. 9.9–9.11): MatLab

p. 89: http://www.math.sfu.ca/histmath/China/1stCenturyAD/NineChapIntro.html

p. 94 (fig. 10.6) http://www-sop.inria.fr/members/Guillaume.Charpiat/publis_en_
    images/faces/AllFinal.png
p. 117: © testing / Shutterstock.com
p. 123: Photo by Eric Draper, http://en.wikipedia.org/wiki/File:George-W-Bush.jpeg
p. 127: © Erik Demaine